The Higgs Boson and Beyond

Sean Carroll, Ph.D.

THE
GREAT
COURSES

PUBLISHED BY:

THE GREAT COURSES
Corporate Headquarters
4840 Westfields Boulevard, Suite 500
Chantilly, Virginia 20151-2299
Phone: 1-800-832-2412
Fax: 703-378-3819
www.thegreatcourses.com

Sean Carroll, Ph.D.
Research Professor of Physics
California Institute of Technology

Professor Sean Carroll is a Research Professor of Physics at the California Institute of Technology. He did his undergraduate work at Villanova University and received his Ph.D. in Astrophysics from Harvard in 1993. His research involves theoretical physics and astrophysics, with a focus on issues in cosmology, field theory, and gravitation.

Prior to arriving at Caltech, Professor Carroll taught and did research at the Massachusetts Institute of Technology; the Kavli Institute for Theoretical Physics at the University of California, Santa Barbara; and the University of Chicago. His major contributions have included models of interactions among dark matter, dark energy, and ordinary matter; alternative theories of gravity; and violations of fundamental symmetries. His current research involves the foundations of quantum mechanics, the physics of inflationary cosmology, and the origin of time asymmetry.

While at MIT, Professor Carroll won the Graduate Student Council Teaching Award for his course on general relativity, the lecture notes of which were expanded into the textbook *Spacetime and Geometry: An Introduction to General Relativity*, published in 2003. In 2006, he received the College of Liberal Arts and Sciences Alumni Medallion from Villanova University, and in 2010, he was elected a Fellow of the American Physical Society.

Professor Carroll is the author of *From Eternity to Here: The Quest for the Ultimate Theory of Time*, a popular book on cosmology and time. His latest book is *The Particle at the End of the Universe*, about the Higgs boson and the Large Hadron Collider.

Professor Carroll is active in education and outreach, having taught more than 200 scientific seminars and colloquia and given more than 50 educational and popular talks. He has written for *Scientific American*, *New Scientist*, *The Wall Street Journal*, and *Discover* magazine.

Professor Carroll's blog, *Cosmic Variance*, is hosted by *Discover*. He has been featured on such television shows as *The Colbert Report*, PBS's *NOVA*, and *Through the Wormhole with Morgan Freeman* and has acted as an informal science consultant for such movies as *Thor* and *TRON: Legacy*.

Professor Carroll's previous Great Courses are *Dark Matter, Dark Energy: The Dark Side of the Universe* and *Mysteries of Modern Physics: Time*. ■

Table of Contents

Table of Contents

The Higgs Boson and Beyond

Scope:

For decades, particle physicists have searched for the elusive Higgs boson particle. In July of 2012, scientists at the Large Hadron Collider (LHC) in Geneva, Switzerland, announced that they found it. In this course, you will investigate what makes the Higgs discovery so important, how the discovery was made, and what implications the discovery has for the future of physics.

The Higgs boson is the capstone of the standard model of particle physics, a comprehensive theory that describes every particle and interaction ever discovered in laboratory experiments. Without the Higgs, the standard model wouldn't work; with it, it successfully accounts for experimental data with exquisite precision.

Understanding why the Higgs is so central begins with an investigation of the underlying principles of quantum field theory. What we perceive as "particles" are really vibrations in fields that pervade all of space. These fields interact with each other in ways that are constrained by underlying symmetries and are elegantly described using the tool of Feynman diagrams.

This course explains how the basic symmetries of the standard model fit together in a way that requires a Higgs field, whose vibrations are seen as Higgs bosons. Unlike other fields, which have zero value in empty space, the Higgs is nonzero everywhere. Particles such as quarks and leptons travel through the Higgs, interacting with it and feeling its influence. It is that influence that gives elementary particles their mass and explains why the weak nuclear force only acts over short distances.

The importance of the Higgs to the world we see is difficult to overstate. Ordinary matter is made of atoms, which consist of electrons orbiting atomic nuclei. If it weren't for the Higgs field, those electrons would have zero mass. According to Einstein's theory of relativity, massless particles always move at the speed of light. In a world where electrons were massless, there would be no atoms, because electrons would never slow down to bind with nuclei. The world of ordinary matter would be a soup of particles zipping by each other; there would be no atoms, no molecules, no life.

The search for the Higgs was carried out at the LHC, which is the largest and most complicated machine ever built by human beings. Staggering in scale and complexity, the LHC was built over the course of years by thousands of dedicated physicists and engineers. The ultimate discovery of the Higgs in 2012 was a testament to their skill and perseverance.

Now that the Higgs boson has been discovered, physicists are optimistic that it's just the beginning. While the standard model correctly describes ordinary matter, that's not all that exists in the universe, which is dominated by dark matter and dark energy. There are good reasons to believe that the Higgs can help us find dark matter particles directly and perhaps help explain the behavior of the dark energy that is making the universe accelerate. Future generations will look at the discovery of the Higgs boson as the crowning achievement for one era in fundamental physics and the beginning of a brand new one. ■

The Importance of the Higgs Boson
Lecture 1

On July 4, 2012, the discovery of the Higgs boson was announced. This was a massive, worldwide media sensation. The Higgs boson is a big deal. Part of the point of this course is not only to explain to you what the Higgs boson is and what role it plays, but also why people were so excited by it. This lecture will explain why the Higgs boson is such a big deal. And lectures to come will go through the details and fill in the story.

The Rise of Particle Physics
- Ancient Greek philosopher Democritus is the father of what we now call "particle physics." And Democritus had the idea that you see a bunch of things around you, including water, solids, and air. And the ancient Greeks would have thought that these were all fundamentally different things. But the atomists—Democritus included—had the idea that, in fact, there was just a small number of fundamental building blocks that go into creating all of these different substances. They called these fundamental building blocks "atoms."

- In the 19th century, chemists started using the word "atoms," so in our modern language, we use "atoms" to speak of the chemical elements. But these "atoms"—the chemical elements—aren't really fundamental. They are made of smaller things.

- What Democritus and his colleagues were really talking about were what we would now call the elementary particles, the fundamental building blocks of nature. And, since the 19th century, physics has been trying very hard to find the elementary particles. Particle physics is a thriving field.

- Atoms are made of elementary particles called electrons. They also have protons and neutrons, which we thought were elementary, but they're not. Protons and neutrons are made of even smaller particles called quarks. In fact, there is a whole zoo of elementary particles, including gluons, neutrinos, and mu-mesons.

- The Higgs boson is a particle. It's a boson, which is one of the two major kinds of particles. But particles are actually not what the world is made of. The idea that is centrally important to modern physics is that the world is made of fields.

- A field, unlike a particle, is spread out everywhere, throughout the universe. It is a number at every point in space and time. A particle has a location, but a field is spread out everywhere and is vibrating, or changing its value. The modern point of view on the fundamental nature of reality is that quantum fields—vibrating at every location, in space, and throughout time—are the fundamental building blocks of nature.

- What we call "particles" are just what we see when we look at the fields. Why do we care that the world is made of fields versus particles? When you look at the field, you see it vibrating. And you see a particle. So, it makes perfect sense to use the language of particle physics to say that there is an electron particle, even though we know there's a vibration in the electron field.

- The one difference is the Higgs boson. The Higgs boson is a particle. It is a vibration in a field called the Higgs field. But, unlike all of the other particles—such as the electron, which is a vibration in the electron field, and the photon, which is a vibration in the electromagnetic field—for the Higgs boson, it's the Higgs field, not the particle, that really matters.

- On July 4, 2012, people were excited to announce the discovery of the Higgs boson. But what got them excited was the knowledge that there is a Higgs field that fills space. Everywhere around you, as you walk through the universe, you are moving through the Higgs field.

- Every particle that we know about is actually a field of one kind or another. But there's a difference between the other fields and the Higgs field. What makes the Higgs special is that, even in empty space, the Higgs field is not zero. Instead, it has a number at every point in space. When there are no electrons around, in empty space, the electron field is said to be zero. All of the fields we know about are sitting at zero, undisturbed—except for the Higgs field.

- The Higgs field plays two crucially important roles in modern particle physics. One role is that it governs the action of the weak nuclear force. There are four fundamental forces in nature: gravity and electromagnetism, which are the strong nuclear forces, and then there are two nuclear forces that only work on very small scales and are called weak nuclear forces. The second main feature of the Higgs boson is that it gives mass to elementary particles.

- If there were no Higgs boson—if there were no field filling space from which the Higgs boson arose—we would live in a very different universe. For example, we would live in a universe where the electron did not have mass. If you had a world where the electron didn't have mass, you would not have atoms, because electrons would not be able to settle into atoms. That means no molecules. That means no chemistry. That means no biology, or no life.

The Standard Model of Particle Physics

- The standard model of particle physics is a complicated theory in particle physics in which all of the forces—gravity, electromagnetism, weak, and strong—have fundamentally different properties.

- The Higgs boson was the last piece of the standard model puzzle. It makes the weak interactions work, gives mass to some of the other elementary particles, and was the last particle that was found. However, because of the theoretical superstructure of the standard model, particle physicists knew that the Higgs boson had to exist before it had been discovered.

- Since the 1960s, particle physicists had thought that there was a field filling space—what we now call the Higgs field—that seemed to make the theories of particle physics make sense. These thinkers, who invented the idea of the Higgs field and the Higgs boson, were driven by trying to understand how nature works. Why does the weak nuclear force have such short range?

- This was a very difficult issue at the time, and it was fundamentally a mathematical problem. We had good ideas about where the weak nuclear force and the strong nuclear force could come from, but they didn't seem to give us the kind of theory that we really wanted. That's why we were eventually driven to this dramatic idea that all of empty space is suffused with this invisible energy field, called the Higgs.

- In the 1960s, or even 1970s, we had the idea of the Higgs boson, but we hadn't actually seen evidence for the field in the data. So, the next step was to find direct evidence for it. In particle physics, that means building new, more energetic, more powerful particle accelerators.

- The way that particle physics gets its direct evidence is by $E = mc^2$, which is Einstein's famous equation relating the energy of an object at rest to its mass, times the speed of light squared. It's the energy that a particle has when it's not moving. This enables us to bring into existence new particles with heavier masses than before, if we can squeeze a lot of energy into a very tiny amount of space.

- A particle accelerator does just that. It accelerates other particles to extremely high velocities. Those particles get a tremendous amount of energy together, and then the accelerator smashes them together, creating more massive particles.

- It was in the 1970s that the search for the Higgs boson began in earnest. The 1970s was an era when quantum field theory became triumphant. We figured out many of the pieces of the standard model of particle physics. We knew that there were quarks, gluons, and W and Z bosons. (We didn't discover the W and Z bosons until the 1980s, but we knew they had to be there before then.)

- The top quark, which was hypothesized in the 1970s, was discovered in 1995 at a particle accelerator called the Tevatron at Fermilab, the flagship United States particle-physics laboratory in Chicago. In addition to looking for the top quark, they were also looking for the Higgs boson.

The Large Hadron Collider

- Just about every major particle accelerator since the 1970s has been looking for the Higgs, but none of them was able to find it. That led us to build the Large Hadron Collider (LHC), which is not only the most impressive particle accelerator ever built, but it is also the largest machine ever built, of any kind, by human beings.

- The LHC is located just outside of Geneva, and even though it's a European collaboration, other countries, such as the United States, are allowed to participate. It is 27 kilometers in circumference, is 100 meters underground, and cost about $9 billion to build.

- Finally, in 2008, we turned on the LHC, and in 2012, we found the Higgs. In fact, we found it sooner than we had expected to find it; the LHC worked better than we thought it would. It was difficult to find the Higgs boson. We knew, from the theoretical construct, how it should interact with other particles, but we didn't know what the mass of the Higgs was supposed to be.

- Nobel Prizes followed for Peter Higgs, who the boson is named after, and François Englert, another one of its discoverers. When you do basic research, looking for a fundamental particle of nature, there will be technological applications. In building the Large Hadron Collider, we have accomplished amazing feats in computer science. In fact, CERN, the home of the LHC, invented the World Wide Web.

Rolf-Dieter Heuer, Director-General of CERN, posing in front of a photo of the CMS detector of the LHC.

- By spending money on basic research, we inevitably get more back, per dollar, then we put in to the science in the first place. But that is not why we do it. The reason we devote money to things like the Higgs boson and the LHC is because we want to discover the way the world works. So far, fortunately, the human race has decided that it's worth the money to look for these fundamental building blocks of nature.

Suggested Reading

Butterworth, *Most Wanted Particle*, Chapter 1.

Carroll, *The Particle at the End of the Universe*, Chapters 1 and 2.

Lederman, *The God Particle*, Chapter 1.

1. If you had lived in ancient Greece, would the idea of atoms have seemed compelling to you? What kinds of arguments might have swayed your opinion?

2. How should modern societies decide how much money to spend on basic research, such as particle physics?

Quantum Field Theory
Lecture 2

In this lecture, you will learn one of the deepest, most profound, and important facts about the fundamental nature of physical reality: The world, at a fundamental level, is made of fields, and the rules of quantum mechanics say that when we look at those fields, we see them as particles. And once you realize that particles are really excitations, or vibrations, in quantum fields, you will understand why particles behave in some of the ways that they do. This is the realm of quantum field theory.

Quantum Field Theory

- The search for fields goes back to Isaac Newton, like many things do in physics. Newton explained that gravity is universal. He came up with a single equation that explained both an apple falling from a tree and the motion of the planets around the Sun. This is Newton's universal law of gravitation, otherwise known as the inverse square law.

The search for the Higgs goes back as far as Newton and his universal law of gravitation.

- Newton's idea was that if you have a gravitating object, such as the Sun, and you have objects around it, such as planets, that object, the Sun, will pull on the planets. That's the force due to gravity. And the strength of that force becomes weaker and weaker as you move farther away from the object. In particular, according to Newton, the strength of the gravitational force dilutes as one over the distance squared. The constant of proportionality is called Newton's constant of gravitation, one of the most important parameters in physics.

- Newton's law, as successful as it is, is also quite mysterious. How do the planets know how strong the gravitational force is supposed to be? This is called action at a distance. And Newton couldn't explain it; he left it to future generations to explain.

- Not long after Newton, around the year 1800, a French mathematician and physicist named Pierre-Simon Laplace came up with a slightly different theory of gravity. Laplace's theory of gravity is basically the same as Newton's theory in all of its experimental predictions, but instead of a gravitational force, there is a field pervading space. There is a number at every point in space, which Laplace called the gravitational potential field. And in Laplace's version of gravity, he replaces Newton's equations with equations for this gravitational potential field.

- Laplace suggested that a big, massive object warps the field. By pushing on the field, it changes its value. In addition, the slope of the field—the amount by which the field is changing—gives you the force due to gravity.

- The answer you get for the force due to gravity is exactly the same in Laplace's view and in Newton's view. But the difference is that Laplace does not have action at a distance. Instead, the field is everywhere, even though you don't see it. So, the Sun affects the field nearby, and that affects the field near that—and so on, throughout the whole universe. And the field is what gives rise to the force of gravity.

- A field is simply a number at every point in space. A particle, on the other hand, has a position. A particle is in one spot and not anywhere else; a field is everywhere. Fields are important because they embody what physicists call locality. In physics, we think that events happen at a place: a point in space and a location in time. Even though fields are spread out throughout space, what happens to a field at any one point is only governed by what's happening to the field at other nearby points. It's not governed by what's happening far away.

- According to field theory, if an influence is going to travel through the universe, it has to start somewhere and affect the field there. And that affects the field nearby. And that's how things propagate through space—just like if you drop up a pebble into a pond, waves will ripple out.

- In Newton's and Laplace's time, we used to think that the world has particles and also fields, but now we've done away with particles. We think that it's all just fields at a fundamental level. The particles are a convenient fiction. Nevertheless, we talk about particles all the time.

Light: Particle or Wave?

- Is light like a particle or a wave? This is a famous physics conundrum. The answer is that it's a wave. But it is complicated. And the complications come up because of quantum mechanics.

- In the 1600s, physicists were having a debate over whether light is a particle or a wave. Dutch physicist Christiaan Huygens argued that light was a wave, because waves have interference and refraction phenomena. This means that it if you send light through a prism, it gets spread out into different colors. Different wavelengths behave differently when they move through different media.

- Likewise, if you pass light through a slit, light can interfere with itself. You can see fringes of interference as the light goes through slits. A wave is up in some places and down in other places. And different waves can add together, either constructively or destructively. If a high wave hits a low wave, you can be left with nothing. Light acts like that in some circumstances, which seems to say that light is a wave in some kind of medium.

- Isaac Newton, on the other hand, argued that light must be particles. Consider the shadows that are cast when light hits an object. Shadows tend to have sharp edges. If you send a water wave around a corner, it bends around the corner. But if you shoot particles at a target, if they don't go past the target, they don't bend around on the other side.

- If light is a wave, then what is actually doing the waving? The answer is the electric and magnetic fields. In the 1800s, when this answer was finally cooked up, we knew about two very important fundamental fields other than the gravitational field: the electric field and the magnetic field. The electric field keeps atoms together; the electric field of the nucleus keeps the electron bound to it.

- When you bring a magnet close to your refrigerator, the refrigerator acts on the magnetic field around it, even though you can't see the field, pulling the two objects together. Michael Faraday figured out how these lines of force behave.

- Danish physicist Hans Christian Ørsted realized that electric fields and magnetic fields are not separate from each other—that they are two aspects of a single underlying phenomenon. Then, in the 1800s, Scottish physicist James Clerk Maxwell showed that he could completely unify our understanding of the electric field and the magnetic field, and now physicists call it a single idea: electromagnetism. Beams of light, X-rays, and radio waves are all electromagnetic waves.

- An electron has an electric field around it. It does not have a magnetic field. But when the electron starts moving, two different things happen. Magnetic fields are created when charged particles move, so once you start shaking the electron, a magnetic field is created that ripples away from the electron.

- In addition, the electric field that the electron has is always pointing toward the electron, but once the electron starts shaking, the location that the electric field is pointing in changes slightly, meaning that there are ripples in the electric field that propagate outward from that electron.

- It's these coupled propagating oscillations—the electric field oscillating up and down and the magnetic field oscillating up and down—that we call an electromagnetic wave, or light. All of the different forms of electromagnetic radiation are electric fields and magnetic fields oscillating in tandem to make a wave.

Quantum Mechanics

- In the 1800s, the idea of fields became known, and they became very popular once we understood electricity and magnetism. Then, they were related to particles with quantum mechanics, which says that what we observe when we look at something is fundamentally different from what really exists when we are not looking at it. That is difficult to believe, which is why there are many interpretations of quantum mechanics.

- In quantum field theory, there are fields, such as the electric field, magnetic field, and gravitational field. But when you look very carefully at those fields, particles are what you see. In particular, they're what you see if the fields are vibrating.

- If you have a field that is not changing, you look at it and see nothing. There's no visible effect. But if you poke it, you start the field vibrating, and then you observe it. It's where those vibrations are that you would say that you saw particles. And every kind of particle is a vibration in some kind of field.

- Where does the correspondence between waves and particles come from? If you have a sufficiently sensitive experimental apparatus (much more sensitive than our eyes), you resolve waves into individual packets of energy called particles. That's not because the wave is individual packets of energy; that's just what we see when we observe it. That is the feature of quantum mechanics, when applied to field theory, that brings particles into reality.

- There are two fundamental kinds of particles: bosons and fermions. Bosons are particles that carry forces, such as the gravitational force, gravitons, the electromagnetic force, and photons. Fermions are the matter particles. They are the electrons, quarks, and neutrinos that make up human beings and even planets.

- For a boson field, the amplitude of the field, or the strength, can be anything. You can make an electromagnetic wave or a gravitational wave of arbitrary size. In the language of particles, we say that the bosons can pile on top of each other.

- Fermions are very different. A fermion particle is a vibration in a fermion field, but the vibrations cannot be arbitrarily big. For a fermion, the field is either vibrating, or it's not. What that means is that fermion particles take up space. At any one point in space, there's either a fermion or there's not. This is known as the Pauli exclusion principle, named after physicist Wolfgang Pauli.

Suggested Reading

Carroll, *The Particle at the End of the Universe*, Chapter 7.

Close, *The Infinity Puzzle*, Chapters 1–3.

Questions to Consider

1. Does it seem sensible to you that quantum mechanics seems to distinguish between what we see and what really is? How much effort should physicists devote to a better understanding of the foundations of quantum mechanics, in contrast with new models of particles and forces?

2. Do you think it's important that, when explaining particle physics to broad audiences, physicists emphasize that everything is really made of fields?

Atoms to Particles
Lecture 3

I f your ultimate goal is to understand the Higgs boson—to know why we know it exists and to find it—then you first need to learn about the other particles of nature included in the standard model of particle physics. There is only a finite number of fundamental particles that we know of, but there could be many more that we haven't found yet. In this lecture, you will be introduced to the various particles that we know exist thus far.

The Basics of Particles

- Every particle has mass, spin, and interactions with other particles. And once you know what those features are for a specific particle, you know which particle it is. The reason every electron has the same mass, spin, and interactions, for example, is because they're all excitations, or vibrations, of a single underlying fundamental field.

- When we talk about the different fields and particles that there are, there is one distinction that is more important than anything else: Is the particle a fermion or a boson? Fermions are the matter particles that make up atoms and structures; they take up space. Bosons can pile on top of each other, so they can lead to a very tangible force in nature. The forces that we know of—gravity, electromagnetism, and the strong and weak forces—are all passed back and forth by bosons.

- The fermions in nature also come in two subsets: quarks and leptons. Quarks are the fermions that feel the strong nuclear force, which is carried by particles called gluons. Leptons are the fermions that don't feel the strong force. In nature, there is an equally big set of quarks and leptons.

- The bosons also come in two families, but there's a big imbalance. Almost all the bosons that we know of are gauge bosons, which are force-carrying particles. The gauge bosons are the particles that carry gravity, electromagnetism, and the weak and strong nuclear forces. The other kind of boson is the Higgs boson. It is all by itself; it is special in many ways.

Particle Spin

- Every particle has an intrinsic spin. Particles have no intrinsic size. This is a fundamentally purely quantum mechanical property that says a zero-sized particle can be spinning.

- Planck's constant is a fundamental quantity of spin. Max Planck was a German physicist who helped pioneer quantum mechanics. He discovered the fundamental unit of quantum mechanicalness, which we use to measure spin.

- For the bosons, the spin is always an integer, such as zero, one, two, three, etc. The Higgs boson has a spin of zero. Almost all of the force-carrying bosons, the gauge bosons, have a spin of one. For example, the photon has a spin of one, and the gluon has a spin of one. The only exception is the graviton, which has a spin of two. Gravity is special, ultimately because of Einstein's great insight that gravity is a feature of space-time itself.

- Fermions, on the other hand, have spins that are integers plus one-half in units of Planck's constant. In fact, usually it's zero plus one-half. Almost all of the fermions, and certainly all of the elementary fermions we've discovered, have spins of exactly one-half. This includes electrons, neutrinos, and quarks. You would need two electrons to make up the spin of one photon.

The Basics of Atoms

- Atoms are the central building blocks of chemistry. Individual atoms make up the chemical elements. But atoms are not fundamental; they are made of smaller particles. At the center of the atom is a hard, dense nucleus. The outside parts of the atoms are where the electrons live, orbiting around the nucleus.

- Even the nuclei aren't fundamental. One nucleus can change into another one. We call this radioactivity. It wasn't until the 20th century that we realized that the reason nuclei can change back and forth to each other—from uranium to lighter elements, for example—is because the nucleus itself is made of smaller particles that we call nucleons, which is a catchall term for protons and neutrons.

- Electric charge is associated with every particle. Protons are the positively charged particles, while electrons are the negatively charged particles.

- Later, another particle was discovered that was almost the same mass as the proton, but it was more difficult to find because it is electrically neutral. This particle is called the neutron.

- The protons and neutrons are held together by some sort of nuclear force to make the nucleus. The electrons are stuck together to the nucleus by electricity and magnetism. The electron, with a negative charge, is attracted to the proton, with a positive charge. This is explained by the fundamental feature of electricity: Opposite charges attract.

- It used to be thought that all nuclei were different. But now we know that they just have different combinations of protons and neutrons. Every nucleus is just a different combination of some number of nucleons.

Energy and Mass
- The mass of a proton is measured in units of electron volts. An electron volt is a unit of energy, not mass. It's the amount of energy it takes to move an electron through one volt of electrical potential.

- If you want to convert from electron volts to everyday energies, one electron volt is a minuscule amount of energy. So, we shift from using electron volts for energy to using giga-electron-volts (GeV), which is a billion electron volts, or about the amount of energy that is in the mass of a single proton.

- A proton and a neutron are approximately equal in mass. The electron is about 1/1800 the mass of a proton, which is about 500,000 electron volts. The Higgs boson has a mass of about 125 GeV.

- There are many other particles, including up quarks and down quarks, that are stuck inside protons and neutrons, so it's more difficult to weigh them. We think that an up quark is about two million electron volts, about four times the mass of an electron. A down quark might be five million electron volts. Neutrinos are very light; they're less than a thousandth of one electron volt.

Matter and Antimatter

- In the 1920s, British physicist Paul Dirac was trying to figure out how electrons behave. He wrote down an equation that turned out to be a very useful one. We now call it the Dirac equation. It is the correct equation for how electrons interact all by themselves.

- But people were skeptical about Dirac's equation because there were two solutions to the equation. One solution describes electrons. The other solution seemed to describe a particle that looks exactly like an electron. It has the same mass and the same spin, but it would be positively charged rather than negatively charged, like an electron.

- In 1928, when Dirac made this equation, no one had ever discovered such a particle. We now know that Dirac's equation was predicting antimatter. At the time, nobody believed it. Then, in 1932, an experimental physicist named Carl Anderson discovered antimatter. He found the antiparticle to the electron, which we now call the positron.

- Not every particle has an antiparticle. For example, photons and gravitons don't have antiparticles. The particles that do have antimatter, such as electrons and fermions, are the ones that carry some conserved number, such as an electrical charge or a fermion number. And the antiparticle always has the negative value of the particle.

- Protons and neutrons, the nucleons, are not fundamental particles. They are made of smaller particles called quarks. For historical reasons, the two kinds of quarks that go into making protons and neutrons are simply called up quarks and down quarks.

Fermions			Bosons
u — Up	c — Charm	t — Top	γ — Photon
d — Down	s — Strange	b — Bottom	Z — Z boson
V_e — Electron neutrino	V_μ — Muon neutrino	V_τ — Tau neutrino	W — W boson
e — Electron	μ — Muon	τ — Tau	g — Gluon
			H — Higgs boson

© Sean Carroll.

- A proton has a charge of +1. A neutron has a charge of zero. An electron has a charge of −1. In the quark model, a proton has two up quarks, with an electrical charge of +2/3 each, and a down quark, with an electrical charge of −1/3—which add up to be +1.

- Quarks cannot be seen in the open. Quarks, and the gluons of the strong nuclear force that hold them together, are confined inside strongly interacting particles. But quarks come in what we call colors. Unlike the leptons or the gauge bosons, quarks can be in red, green, or blue. Then, antiquarks are the anti-colors: anti-red (cyan), anti-green (magenta), and anti-blue (yellow).

- The trick to understanding quarks is that we only ever see colorless combinations. That means that either quarks are in groups of three—one red, one green, one blue, which is a proton or a neutron—or they are in groups of one quark and one antiquark, such as a red and an anti-red, which is a meson.

Particle Interactions
- If you take a neutron out of an atomic nucleus and let it be there in empty space, it will decay after about 10 minutes, on average. Neutrons are not stable particles. But if you put a neutron in empty space outside the nucleus, that neutron will decay into a proton, an electron, and an antineutrino.

- Every time there's a change from one kind of particle configuration to another, that means there is a force at work. The electromagnetic force doesn't change the identity of particles, nor does gravity or the strong nuclear force. These forces push around the particles, but they don't change them.

- The weak nuclear force is the force of nature that can alter the identity of fundamental particles. When a neutron gets altered into a proton, an electron, and an antineutrino, we now know that it's an up quark converting into a down quark, or vice versa.

- Of all the forces of nature, the weak force is the most complicated—and this is because of the influence of the Higgs boson. There are three kinds of bosons that carry the weak nuclear force. There is a Z boson, which is neutral and weighs about 91 billion electron volts (91 GeV). There are two W bosons, which are electrically charged. There's a positively charged W boson (W^+) and a negatively charged W boson (W^-). These weigh about 80 GeV.

- It's the W boson that can change one particle into another one. An up quark, which has a charge of $+2/3$, can emit a positively charged W boson and turn into a down quark that has a charge of $-1/3$. Electric charge is conserved in all of these interactions.

- The W and Z bosons decay very quickly. You can never see a W boson or a Z boson in a particle physics experiment; all you can do is see what they decay into. But it turns out that that's a very common thing in modern particle physics. The same is true of the Higgs boson.

- In 1936, Carl Anderson found evidence for another kind of particle that was heavier than the electron but had the same charge. He realized that this new particle, which he called the muon, was exactly like the electron but heavier. This didn't fit in with anything we knew or had predicted about particle physics.

- The discovery of the muon opened a new window on how particles work. We now realize that the four fermions—the up quark, the down quark, the electron, and the neutrino—are a family, or a generation, of particles. In fact, there are three families of fermions. In each family, there are two quarks and two leptons. Of the two leptons, one is charged and one is neutral (the neutrino).

- The electron has a heavier cousin called the muon, which has a heavier cousin called the tau. The neutrino is now called the electron neutrino, and there is also a muon neutrino and a tau neutrino. There's an up quark and a down quark. But there's also a charm quark and a strange quark. And there's a top quark and a bottom quark.

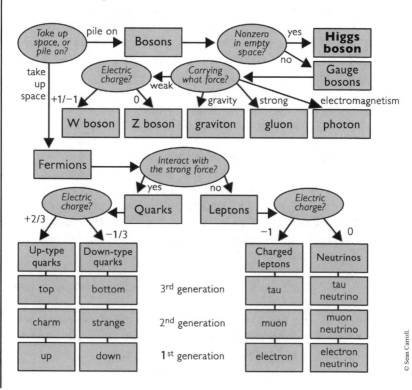

© Sean Carroll.

Suggested Reading

Carroll, *The Particle at the End of the Universe*, Chapter 3.

Lederman, *The God Particle*, Chapter 5.

Questions to Consider

1. Does the standard model seem remarkably compact and elegant or unfortunately sprawling and untidy?

2. Do you think physics will ever achieve a complete understanding of all the particles and forces of nature?

The Power of Symmetry
Lecture 4

W hy did physicists believe that there was a particle called the Higgs boson long before they had any evidence for its existence? To understand what gave them that confidence, you need to learn some profound facts about the essential nature of physical reality. In this lecture, you will learn about symmetry, which plays a fundamentally important role in modern physics. In fact, the underlying structure of the forces of nature is ultimately due to symmetry.

Symmetry

- A symmetry is a change that doesn't make a difference. It's something you can do to something without changing how it appears or how it behaves to you. Examples of flipping and rotating geometric designs and spheres are symmetries of space, because we're physically moving things around.

- There are many symmetries of space that are lurking in the laws of physics.
  - There are spatial translations, which involve moving something from one place to another place and it's still the same thing.

  - In addition, there are time translations. If you have an object that is not changing in time, that's a way of saying that it's the same now as it is at some other moment.

  - Finally, there are boosts, which imply that it doesn't matter how fast we are moving. If you do a physics experiment in a room, you can do the same physics experiment in a rocket ship zooming through space at a constant speed and get the same answers.

- If an experimental outcome doesn't depend on whether you rotate it, move it, or do it at some other time, then there is a symmetry underlying that lack of difference. For example, if you measure the charge of the electron, you will get the same answer everywhere on Earth at every other moment in the history of the universe. There is a symmetry that is keeping the charge of the electron constant.

- A global symmetry is a symmetry that we can do to the whole universe. But we have to do it everywhere all at once. Imagine rotating the universe all at once. That is a global symmetry. On the other hand, a local symmetry, or gauge symmetry, is a symmetry that we can do in different ways, independently, everywhere throughout the universe. Imagine, for example, that you are doing an experiment to measure the charge of the electron. There is a rotational symmetry, so you can do the experiment, rotate the whole laboratory by any angle, do the experiment again, and get the same answer.

- In fact, there's a much larger symmetry involved because there is a local symmetry. Imagine that there are two laboratories, one of which is measuring the charge of the electron while the other is measuring the charge of the proton. Not only can we rotate the world, or some big platform that both of those laboratories are on, but we also can rotate one laboratory with respect to the other one and the two experiments will still get the same answer.

- When you have a lot of symmetry, that is restricting what can happen. In terms of geometric designs, a circle is more restrictive than a square, which is more restrictive than a rectangle. The fact that you have a symmetry that you can do separately everywhere in the universe—local, or gauge, symmetry—puts an incredibly strong restriction on the laws of physics.

Connection Fields
- How do we know that we've done a transformation? How do we know that one lab is rotated, or moved, or translated in time compared to the other one? In real life, we put up some measuring sticks or use some surveying equipment. But the only reason we can

do that is because there is a field that is connecting the orientation of one lab to the other one. In other words, there's a field filling space that lets us compare the orientations of the locations of different parts of the universe. That field is called a connection.

- A connection field is something that lets you compare the outcomes of different symmetry transformations at different parts of the universe. So, every time we say that there is a symmetry that you can do separately in two locations, there must be an underlying connection field that lets you do that. In the example of the laboratories, which are on Earth sitting on the ground, the ground itself is acting as a connection field.

- We notice connection fields filling the universe when they're altering from place to place. When you have a connection field that is changing from place to place in the universe, you feel a force. For example, there's a connection field pervading space-time, and if that connection field is curved, or changing from place to place, you feel a force—one called gravity.

- There are many more connection fields other than the one that gives rise to gravity. The difference is that for the other kinds of connection fields filling space, some particles feel them and others don't. But gravity is felt by absolutely everything.

- Protons and neutrons are very different; they have different electrical charges. But they act more or less the same if you ignore the fact that their charges are different. They have about the same mass, and they behave very similarly with respect to the strong nuclear force. And the electromagnetic force is not that strong when you compare it to the strong nuclear force.

- So, there's an approximation that says that we can treat the proton and the neutron as the same. If we only care about the strong nuclear force, the proton and neutron can be switched with each other, and you would get the same answer. That is a symmetry.

- In other words, you could switch up quarks and down quarks. And because they're both much lighter than the protons and neutrons themselves, you get the same answer. There is a symmetry that is relating the up quark and the down quark, or the proton and the neutron.

- The problem is that even though it's a useful approximation, it's still an approximation. Protons and neutrons are not precisely identical particles, nor are up quarks and down quarks.

- A gauge symmetry, or local symmetry, is something that you can do everywhere in the universe independently. That's very restrictive. One of the restrictions is that gauge symmetries are always exact. They're either true, or they're not true. They're either real, or they're not real.

- A global symmetry is what is between protons and neutrons. It's a useful calculation that approximates neutrons and protons as the same. But it's not really a gauge symmetry.

- Individual quarks have a very noticeable gauge symmetry between them. An up quark is not symmetric with respect to a down quark. They have different masses, so they can't be identical—they can't be switched for each other.

- You can take every quark everywhere in the universe—it's either red, green, blue, or some combination—and you can rotate them in color space, in a space where the three axes are redness, greenness, and blueness. And you can do this separately, independently, everywhere throughout the universe. That's a perfectly good symmetry.

- What that means is there must be a connection field that talks to the colors, a connection in color space. And that connection can change from place to place. And that gives rise to a force: the strong nuclear force, which is sometimes called quantum chromodynamics. It's the dynamics of color itself, with color being the symmetry that this gauge connection field is talking to.

- All of the four forces of nature are gauge forces. These forces all arise from an underlying symmetry that is throughout the world, local and independent, and therefore comes with a connection field filling all of space.

- Quarks feel the quantum chromodynamic strong nuclear force. The fermions that we know about feel the weak nuclear force. And everything feels gravity. But you can think of all of empty space as being alive with small tilted slopes that cause the forces that push around particles through the universe.

Symmetries and the Forces of Nature
- Whenever you have a gauge symmetry, you have a connection field. Whenever you have a connection field, there's a particle. These particles are the gauge bosons, the particles that we say carry the forces of nature.

- The idea that the forces of nature are based on symmetries is a profound one. Of course, it was true for gravity and electromagnetism, but the suggestion that it was true more generally goes back to 1954. Yang and Mills, two physicists, proposed that all forces of nature are based on these gauge symmetries. This is Yang-Mills theory, and it is a quantum field theory.

- The idea of these symmetries underlying all the forces of nature was not necessary when we were thinking about gravity and electromagnetism. We already thought that we understood those forces before Yang and Mills came along. What was mysterious were the nuclear forces, what we now call the strong and weak nuclear forces. Can we understand the nuclear forces as Yang-Mills theories, based on gauge symmetries and their associated connection?

- Eventually, the Higgs boson was realized as the solution to this problem. In 1957, physicist Julian Schwinger, who shared the Nobel Prize with Richard Feynman for understanding electromagnetism, was trying to understand the weak nuclear force as a Yang-Mills theory.

- Schwinger's theory predicted two massive electrically charged gauge bosons, and he called them the W^+ and the W^-. This is the origin of what we now call the W bosons. He predicted one other particle that was neutral. He did not give it a mass. He called it the photon.

- In other words, Schwinger was proposing that there were these three symmetries of nature. Two of them became massive gauge bosons and were the weak interactions. One of them remained massless and was electromagnetism. He was unifying these different forces of nature, something that physicists have tried to do ever since.

- Unfortunately, Schwinger's idea was a complete failure, not only because he was cheating by giving the W bosons a mass by hand, but also because the weak interactions and the electromagnetic interactions are very different. The difference is something called parity violation.

- Parity is a kind of symmetry. A mirror exchanges left with right; it's doing a parity transformation, inside and out. For example, if a particle is spinning clockwise, under parity it switches to spinning counterclockwise.

- It turns out that the strong, the electromagnetic, and the gravitational forces are invariant under parity. It doesn't matter whether you're in our world or in the mirror-image world. But the weak interactions violate parity. The weak interactions know if you're in the real world or in the mirror image. Somehow, the weak interactions know the difference between clockwise and counterclockwise.

- Schwinger passed this problem on to a bright young student he had named Sheldon Glashow, who was stuck with unifying two forces that had very different properties, electromagnetism and the weak force.

- Glashow's theory was an elaborate, complicated set of symmetries, but it predicted four gauge bosons: two W bosons, a Z boson, and a photon. The symmetries that Glashow was suggesting were not

- A short-ranged force doesn't obey an inverse square law; it falls off in strength much faster than a long-ranged force. Typically, a short-ranged force falls off exponentially fast, which means that instead of 1 over the distance squared, the strength of the force goes as some number to the power of minus the distance ($e^{-r/L}$).

- It makes sense to have an inverse square law for forces. There are lines of forces of gravity coming from the Sun, and they get diluted as they spread over spheres at larger and larger distances. Mathematically, in quantum field theory, what that implies is that a long-range force is mediated by a massless particle. Gravitons, which are the hypothetical particles of gravity, have zero mass. When you have a long-range force, you have massless particles, and vice versa.

- Gauge symmetries have connection fields that tell you how to relate the symmetry transformations from one point to another, and it is a mathematical fact that bosons—the particles that are associated with those connection fields—are always massless. So, we expect that gauge symmetries give rise to long-range forces. But why are the nuclear forces not long-range forces?

- A possible answer came about in the 1950s and 1960s with the idea of breaking symmetries. You can have a symmetry that is underlying your equations, but then you can posit that perhaps there's a field that fills all of space. And the existence of this field breaks the symmetry. There were a number of physicists who suggested this idea, including Yoichiro Nambu, who won the Nobel Prize for this idea in 2008.

- Maybe the weak and strong nuclear forces are gauge theories; maybe they are associated with massless bosons. But maybe those symmetries are broken. And breaking the symmetry gives a mass to the bosons. If the math says that symmetry implies masslessness, then maybe breaking the symmetry gives you a mass. And that would explain the short-range forces.

- Alas, it's not that simple. British physicist Jeffrey Goldstone proved a theorem that says that you can break symmetries by putting a field in empty space. But then even if you give masses to the gauge bosons, they are a new kind of boson that is invented, which we now call Goldstone bosons, which are also massless.

- Goldstone's theorem seemed to imply that whenever you had a symmetry, you would get a massless particle of one form or another. And massless particles are easy to make in a particle accelerator. The conclusion seemed to be that the nuclear forces could not be Yang-Mills theories, theories based on gauge symmetries, because those theories would necessarily lead to massless particles. And we didn't see such massless particles.

- In fact, that logic was wrong. Gluons that carry the strong nuclear force, and gluons are massless, just like the theorems postulated. The difference that people hadn't thought about is that the strong nuclear force is strong. Not only does it have a very strong interaction strength, but the gluons interact with each other.

- Photons of electromagnetism interact with electrically charged particles, but the photons themselves are not electrically charged. However, gluons interact with colored particles, such as quarks, and the gluons themselves are colored.

- This leads to a phenomenon called confinement in the strong interactions. The gluons that are inside a proton or a neutron are interacting with each other. When you're outside the proton, there's almost no evidence of the strong nuclear force, but if you could put your head inside a proton, it would be very, very strong.

- It took years to figure this out, and it made people very skeptical about the very idea of quarks because you couldn't actually see them. It turns out that they are confined inside the proton and the neutron. So, the gluons, the particles that are carrying the force, are massless. But we don't notice because the gluons never escape to the outside world.

unify the weak interactions with electromagnetism when the weak interactions violate parity and electromagnetism doesn't?

- Weinberg realized that if he added the Higgs mechanism to his theory, it would not only give mass to the W and Z bosons and the gauge bosons of the theory, but also this Higgs field in empty space would interact with the fermions. The influence that the Higgs in the background would have on the electron, the muon, the tau, etc., would be to give them mass.

- Imagine a would-be massless electron moving through space. If there were no Higgs there, it would be moving at the speed of light—it would be massless. But because the electron keeps feeling the influence of the Higgs field all around it, that is what gives it mass.

- So, Weinberg realized that adding the Higgs field to his model of the weak interactions gave him an unexpected bonus consequence that he could have a theory that violates parity but gives mass to all of the fermions. And that's how we think fermions actually get their mass in the real world.

- This brilliant idea has influenced particle physicists ever since. Weinberg's theory of leptons not only explains the weak interactions being short range but also gives mass to the electron and to the other particles. Furthermore, the amount of mass is not arbitrary; it's proportional to the strength of the interaction with the Higgs field.

Suggested Reading

Carroll, *The Particle at the End of the Universe*, Chapter 11.

Close, *The Infinity Puzzle*, Chapters 8–9.

Giudice, *A Zeptospace Odyssey*, Chapter 8.

1. If the Higgs field fills empty space, do you think it's likely that there are other fields that act in a similar way that we haven't yet thought of or discovered?

2. How important is it that the true history of how an idea was developed is accurately conveyed when we talk about the idea itself?

Mass and Energy
Lecture 6

I f there is an invisible field filling space—the Higgs field—then we can explain why the weak interactions are short range, which is equivalent to saying why the W and Z bosons have a mass, and why the fermions of the standard model of particle physics also have a mass, even though the weak interactions violate parity. Clearly, the Higgs boson plays an important role in giving mass to the particles of the standard model. In this lecture, you will discover what is meant by the word "mass" and how it relates to energy.

Mass

- Of course, mass is related to the concept of energy: $E = mc^2$. But what does it mean to talk about the mass of a field? The way to think about mass is as the resistance you feel when you try to push or move something. The inertia that something has is proportional to its mass.

- Newton's second law of physics is force equals mass times acceleration ($F = ma$). This is the definition of what we mean by mass. You put a force onto something, and it accelerates. The mass is proportional to the amount of force you need to give some required acceleration. If the mass is high, then a given amount of force doesn't get you much acceleration; if the mass is low, the same amount of force will accelerate you quite a bit.

- Mass is an intrinsic property of some particle or object. Your mass would be the same even if you didn't have any gravity around you, even if you were far away from any gravitational field.

- Inertial mass is the resistance to acceleration, but the gravitational mass of an object is how much gravity an object causes. In our world, the inertial mass equals the gravitational mass. This was first discovered by Isaac Newton.

- This is called the principle of equivalence in Einstein's language. Einstein used the equality of inertial and gravitational mass as a starting point for his development of general relativity, his great theory of gravity as curved space-time.

- One way of thinking about the principal of equivalence is if you drop two objects, if there's no air resistance and if you drop them in the same gravitational field, they will fall at the same rate. It doesn't matter how much they weigh or what they're made of.

- That's very different than charged particles in an electric field. A positively charged particle and a negatively charged particle will move in different directions in an electric field. Electric forces act differently depending on charge. But gravity is universal and, thus, a consequence of the principle of equivalence.

- But that's a feature of the force of gravity, not a feature of the concept of mass. In a world without gravity, there would still be mass. There would still be the resistance to being accelerated when you push something. So, the Higgs gives inertial mass to particles. It has nothing to do with gravity.

Energy

- Einstein's theory of special relativity says that space and time are related to each other. It also says that you can't travel faster than the speed of light. Mass does not increase as you travel faster and closer to the speed of light; independent of who is measuring it and how fast they are moving. But your energy does increase—in fact, it approaches infinity.

- If an object that is moving very fast has a lot of energy, what about as the object moves more and more slowly? Einstein realized that even if an object is not moving, it still has energy. That's the origin of $E = mc^2$: The energy of an object that is not moving is equal to its mass times the speed of light squared. The mass is intrinsic to an object, and that's proportional to its total energy.

- There are various different forms of energy that an object can have. It can have rest energy, which is represented by $E = mc^2$. It can also have kinetic energy, or energy of motion, which is the kind of energy that goes to infinity, according to Einstein, as you travel closer to the speed of light. Finally, an object can have potential energy, which depends on where the object is. Potential energy gets converted into kinetic energy.

- Energy as a whole is conserved in particle physics. But it can be converted from one form to another. Mass is the energy you have when you're just sitting there. Therefore, mass is not conserved. But mass is just one form of energy, and it can be converted into other forms.

- In the standard model of particle physics, the elementary particles that have mass—including quarks, leptons, and gauge bosons—get their mass from the Higgs boson. But mass does not come from the Higgs boson, meaning that objects do not get their mass by summing up the masses of the elementary particles. People get most of their energy from protons and neutrons from the strong interactions, not the Higgs field.

- But what about fields themselves? Mass is the energy that an object has when it's at rest. The problem is that a field isn't an object, and it's not localized in some position. Fields are everywhere. So, a field can't have a mass. You can't associate $E = mc^2$ with a field because a field cannot sit at rest. The particle excitation of a field has mass, but the field itself doesn't have mass in any ordinary sense.

- What the field does have is energy. A field has a value at every point in space and time. And just as particles can have kinetic energy, potential energy, and rest energy, fields can have kinetic energy (how fast a field is oscillating or changing through time) and potential energy (the intrinsic amount of energy the field has, even if it's unchanging). In addition, the new thing that a field can have that a particle can't have is gradient energy, which means that the field can change from place to place and that those changes require energy.

- Every particle is the vibration in a field. But when you start a field vibrating, there will be kinetic energy in the field because it's moving. And there's gradient energy because it's moving differently from place to place. The mass of the particle, however, is associated with the potential energy of the field.

- So, some fields simply have zero potential energy. It doesn't matter what value they have; they have the same amount of energy. When you poke those fields and start them vibrating, you get a particle. And the mass of that particle is zero. So, when you have a field with zero potential energy, its particle excitations will be massless.

How Energy Evolves through Time

- Energy likes to spread out and dissipate so that at any one point in space, energy is minimized. If you put a ball on a hill, it will roll to the bottom of the hill, roll around, and eventually settle down. The total amount of energy is conserved, so when the ball is rolling down the hill, it needs to dissipate its energy. It does this by spreading out.

- What this means for particles is that heavy particles like to decay. The Higgs boson is a very heavy particle. It's a vibration in a field, with a lot of potential energy. But it can convert into many lighter particles. And that's what energy likes to do. That's why the heavier particles, such as the top quark or the Higgs boson, quickly decay. And that makes them very difficult to find.

- What makes the Higgs field special is that it's not zero, even in empty space. But why does the Higgs field, unlike every other field, want to be nonzero in empty space? The answer is that being nonzero is what minimizes the energy contained in the Higgs field. The Higgs, unlike everything else, would have more potential energy if it were stuck at zero than if it were moved out to some large value.

- Most fields that have potential energy will have a potential that its lowest point sits at zero. Most fields in empty space want to be sitting at zero. That's what minimizes their energy. But the

Higgs field is different. At zero value, it has a local maximum of its potential energy—like it's sitting near the top of a hill. The minimum of its potential energy is at a nonzero value: 246 billion electron volts (GeV). That's where the Higgs wants to be; that's its minimum energy state.

- The energy density in the Higgs field is minimized at the bottom of its potential, and it's a huge difference in energy. Imagine that you take a ball the size of a Ping-Pong ball and outside that ball, the Higgs field does what it normally does, but inside that ball, you push the Higgs field back to zero from its nonzero value. That means there's a tremendous amount of energy in that ball. One Ping-Pong ball size of space, with zero Higgs field, would have the same energy as the entire Earth.

- The Higgs field wants to live at a nonzero value, and that zero is special. It is a symmetric value. For example, you can multiply zero by −1, and you get zero back. If the field were at zero, it would be symmetric. If it's not at zero, it is breaking the symmetry. That is why the Higgs field in empty space breaks the symmetry, which is crucial to making the weak interactions work.

- If you change the value of the Higgs field from its nonzero value back to zero, it costs you energy. But this is a statement about the difference in energy, from the bottom of the potential to the hilltop. It's not a statement about the total amount of energy at either point.

- In particle physics, the total amount of energy in the potential of a field is an arbitrary number. It is something you have to measure. If you put the Higgs field at zero energy, or zero field value, then the fact that the Higgs field is a lower energy at a nonzero value means that the energy there is negative.

- We don't think that empty space has this gigantic negative energy, but it's something you could imagine being true. In fact, we believe that measurements have shown us that empty space does have energy. It is what we call the cosmological constant, or the dark energy.

- The cosmological measurements of galaxies accelerating away from each other seem to indicate that the energy of empty space is positive. It is a very small positive number. Why is the energy of empty space small and positive, even though the contribution of the Higgs field seems to be large and negative?

- The answer seems to be that there are some other fields that are canceling the Higgs field. When you take all the fields in nature and add their contributions to the energy of empty space, all the contributions are big. But some of them are positive and some of them are negative. And they almost exactly cancel out. And nobody knows why. This is a fundamental puzzle for physics that is called the cosmological constant problem.

Suggested Reading

Butterworth, *Most Wanted Particle*, Chapter 3.

Randall, *Knocking on Heaven's Door*, Chapter 1.

Questions to Consider

1. Why does Einstein's equation $E = mc^2$ have such powerful resonance in the popular imagination?

2. How do different forms of energy (kinetic, potential, gradient) transform into each other over the course of your everyday life? How is energy being transferred between different forms when you work out at the gym or accelerate your car?

Colliding Particles
Lecture 7

B
efore we start building particle accelerators to look for the Higgs
boson, we have to know exactly what we're looking for. It's not
enough to simply know what particles are in the standard model; we
have to know how they interact. What is the rate of production of the Higgs
boson? And then how does it decay? Instead of doing detailed mathematical
calculations, you can use Feynman diagrams to calculate these rates. As you
will learn in this lecture, these diagrams get complicated, but they provide
insight into the fundamental processes of our universe.

Feynman Diagrams
- In the 1940s and 1950s, physicist Richard Feynman was faced
  with the task of calculating the rates of different particle physics
  interactions. For the purposes of these calculations, Feynman figured
  out how to answer questions in quantum field theory and particle
  physics by using diagrams with lines representing the positions of
  particles. Feynman diagrams are used both to see what can possibly
  happen and to calculate the rate at which it can happen.

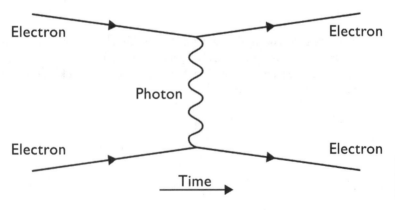

© Sean Carroll.

- These diagrams are simply shorthand for what can happen between different particles, and each different kind of particle is represented by a different kind of line in the Feynman diagram. The force-carrying gauge bosons—including the photon, the graviton, the W and Z, and the gluons—are all represented by wavy lines.

- The Higgs boson is not a gauge boson. In particular, the spin of the Higgs is zero. The spin of the gauge bosons are all one, except for the graviton, which is spin two. A spin zero boson is called a scalar boson. Any particle that is spin zero is a scalar boson. In a Feynman diagram, a scalar boson is represented by a single unadorned straight line.

- Fermions are also straight lines, but they have a little arrow attached to them. And that's because, unlike the Higgs, which is a single particle all by itself, fermions have antiparticles. So, we attach a little arrow to the fermion lines to represent the difference between a particle—such as an electron, a muon, a tau, a neutrino, or a quark—and an antiparticle, such as the positron, the anti-muon, and so on.

- If our Feynman diagrams move left to right, then a particle going left to right has an arrow pointing from left to right. An antiparticle moving from left to right has an arrow pointing from right to left; the direction in which the antiparticle is moving is opposite the way we draw its arrow.

- Finally, we have not only the individual lines, but we put the lines together. This is when we get the interactions between the different particles. Most of the interactions in particle physics only have three particles involved: Either one comes in and two go out, or two come in and one goes out. These are called vertices, which are tricornered places where one line comes in, another one comes in, and the third one intersects both of them. Every such vertex is attached to a number, the coupling constant, which is proportional to the interaction strength.

- For every interaction the Higgs has with a fermion in the standard model, there's a separate coupling constant. And that coupling constant determines the mass of that particle. But it also helps us calculate the rate of decay of the Higgs into other particles.

- If you have a diagram where one particle comes in and two go out, you automatically have other diagrams that are obtained from that diagram by moving the lines around. You can take any line that is on the left side and rotate it over to the right side, and you can take any line that is on the right and rotate it over to the left.

- There's a symmetry deep in the heart of quantum field theory called crossing symmetry. Once we know the number attached to any one of those diagrams, it's the same number attached to any of the other diagrams we get by rotating it. Then, we can rotate further.

- If one Higgs boson can decay into a quark and an antiquark, it is also true that a quark can emit a Higgs boson. And not only can we draw that diagram, but the number we attach to it is the same. By rotating the diagram, we have a quark and an antiquark coming together and annihilating to make a Higgs boson. Finally, we could have a quark, or an antiquark, absorb a Higgs boson.

- Feynman diagrams are magically flexible, and once you draw the simple ones, many implications follow. However, it's not true that you can just draw any Feynman diagram. Some of them exist, and some don't.

- When thinking about which ones do and don't exist, there are two rules to keep in mind. First, there are quantities that are conserved in particle physics interactions, such as electric charge, which is neither created nor destroyed. This means that at every vertex in the Feynman diagram, the same amount of charge comes in and goes out.

- Another quantity that is conserved is called a fermion number, which is calculated by taking the number of fermion particles and subtracting the number of antiparticles. A crucial consequence of this feature is that fermion lines never end. Think about these conserved quantities as flowing through the diagram.

- The second subtlety to keep in mind is that we can draw diagrams that obey the rules, but they don't necessarily correspond to real processes. For example, one of the basic fundamental processes in particle physics is an electron and a positron can annihilate into a photon, or a photon can convert into an electron and a positron. These are perfectly good diagrams, but the processes never happen by themselves, ultimately because of energy.

Virtual Particles

- In quantum theory, we have not only real particles that can be observed, but we also have virtual particles. The definition of a virtual particle is simply a particle that only appears on the inside of a Feynman diagram, never on the outside.

- Virtual particles have a mass that is different than the mass that the particle would have if it were a real particle rather than a virtual one. Inside the Feynman diagram, virtual particles can have any mass at all. This is true because these aren't really particles; they are actually vibrations in quantum fields. We only associate mass with them when they become real particles later.

- So, when we're putting virtual particles inside our Feynman diagrams, they don't need to have the masses that the particles usually have. However, the whole point of the Feynman diagram is to calculate a number, the rate at which this interaction happens. It turns out that if the particles inside the diagram have masses that are close to the masses that the real particles would have, the contribution from that diagram will be larger.

- Another important feature of virtual particles is that they let us have loops inside Feynman diagrams. For example, an electron and a positron scatter by making a photon, and then they go their separate ways. They could also scatter by emitting two photons. And we sum up all of the possibilities, which means that for every diagram where there's a loop, you sum up every possible way that energy and momentum can flow through the lines in the loop. In mathematical terms, that means you're doing calculus, integrating all the different possible ways that momentum can move through the diagram. Calculating the actual effect of loop diagrams is very difficult.

Real Diagrams
- In the standard model of particle physics, some diagrams exist, and some don't. You always need bosons involved in standard model diagrams, so when listing all of the possible diagrams, you make a list of all the bosons and ask what they interact with. For example, bosons can interact with each other and with fermions, gluons can interact with gluons, Higgs bosons can interact with Higgs bosons, etc.

- Think about all the bosons in the standard model and ask what they do. The first one is the photon, the particle of electromagnetism. Photons interact with particles that have electric charge. They do not interact directly with particles that are neutral. So, you can have an electron spit off a photon, for example. You will never have a Higgs boson directly interact with a photon.

- Next are gluons for the strong interactions. Just as photons interact with charged particles, gluons interact with colored particles. And the colored particles are simply the quarks and the gluons. So, a down quark can emit a gluon. It stays a down quark, but its color changes. A Higgs boson cannot emit a gluon because Higgs bosons have no color.

- The Z boson is part of the weak interactions, along with the W boson. The Z boson is a neutral, fairly simple particle. The Z boson interacts in a way similar to how a photon does. But instead of interacting with charged particles, it interacts with every particle that feels the weak interactions, including the fermions, the quarks, the leptons, the Z and W bosons, and even the Higgs boson.

- The W bosons arc the other particles that carry the weak interactions. But they have a crucial difference from the Z bosons, because the W bosons are electrically charged (W^- and W^+). The W bosons are the only gauge bosons that have electric charge. When a particle emits or absorbs a W boson, it changes its identity. The Higgs boson plays a role in the weak interactions; it is neutral, but it can decay into two W bosons.

- Gravitons are particles that many people wouldn't even count as part of the standard model because we don't truly have a quantum theory of gravity. But the rule is easy: Gravitons interact with everything. Every particle can emit gravitons. The Higgs boson can emit or absorb gravitons, and the gravitons can emit gravitons themselves. Gravity is universal, which is part of the reason why it is different than all of the other forces.

- The Higgs boson only interacts with massive particles. If a particle has mass, such as quarks or the W and Z bosons, the Higgs interacts with it. If a particle is massless, such as the photon or the gluon, then there is no direct interaction between that particle and the Higgs boson. Heavy particles interact with the Higgs very easily; light particles interact with the Higgs but more reluctantly.

Suggested Reading

Carroll, *The Particle at the End of the Universe*, Appendix 3.

Kane, *The Particle Garden*, Chapter 4.

Strassler, *Of Particular Significance*.

1. Given the known particles and interactions, can you think of different promising ways that the Higgs boson could be created or detected?

2. Is it an accident or a profound feature of reality that we can get a good approximation by just looking at a few Feynman diagrams, rather than summing up an infinite series?

Particle Accelerators and Detectors
Lecture 8

W e used to be able to find new particles in our environment, but today we have to make them. We have to build a gigantic particle accelerator to create the elementary constituents of matter that are not available in our environment. As a result, the process of designing and building particle accelerators is immensely important. In this lecture, you will not only learn about the particle accelerators that we have and what they do, but you also will learn the logic behind particle acceleration.

The Making of New Particles

- The basic idea behind accelerating particles and making new ones is $E = mc^2$: The energy of an object at rest is its mass times the speed of light squared. If you want to create a massive particle, such as the Higgs boson, you need to put a tremendous amount of energy into a very tiny region, get those quantum fields vibrating, and make that into the new particle.

- The way we create a tremendous amount of energy in a very small region is to smash particles together at high velocity. That's why most particles that we discover today are done so at particle colliders.

- It used to be that we would just collide fast-moving particles into stationary targets. Once the technology developed to the point where we could set two beams of particles against each other, we gained a tremendous reach in discovering new things.

- Particle colliders bring into existence new particles that weren't there. If we smash two protons together in the Large Hadron Collider (LHC) and make a Higgs boson, there were no Higgs bosons hiding in those protons. We are creating new particles by getting the quantum fields they're associating with vibrating.

- What we're really doing is smashing together fields, not particles. When we have two protons coming together and we watch what they make when they collide, the protons themselves are collections of quarks and gluons, and every quark and every gluon is a vibration in the quark fields or the gluon fields. The interactions between particles are vibrations in one kind of field talking to the vibrations in the other kind of field.

- There are three things that we need to do to make new particles, whether it's the Higgs boson, or the top quark, or something else. First, we need to accelerate the particles that we have to as high of an energy level as we can. Because $E = mc^2$ and because we found the low-mass particles already, the particles that we're looking for tend to be those that have a large amount of mass. So, we want as much energy as we can possibly squeeze into a tiny volume that will let us reach these new kinds of particles.

- Second, once we've accelerated the particles into tremendous energies and brought them into collisions, those particles are going to interact with each other, and new particles will come out. We want to look at what comes out. This is the difficult part. There's a whole art and science of looking at what comes out of the particle collisions and reconstructing what must have gone into making that happen.

- Finally, we need to compare the data that we get from an accelerator to our theoretical expectations by analyzing Feynman diagrams and interactions and by calculating the numbers. The data that comes from a particle accelerator is a mess. Unless you know exactly what you're looking for, it can be very difficult to pick out the very faint signal of a new particle. Experimentalists and theorists need to work together to make these particle accelerators do their job.

Accelerating Particles

- Imagine that we are designing a new particle physics experiment. We're going to collide particles together to make something new. What kinds of particles should we be colliding together? Because of the standard model of particle physics, we know what particles there are that we have access to right now—including the leptons, the quarks, and the gauge bosons.

- In terms of the leptons, we have the charged leptons and the neutrinos. Quarks are confined inside strongly interacting particles, so we don't have access to individual quarks. The generic name for a strongly interacting particle like a proton—which is three quarks, or a pion, which is a quark and an antiquark—is hadrons. Hadrons are any collection of quarks and gluons that are overall something that has no color, something we can make in the laboratory.

- What would make a good candidate? We want a particle to accelerate that is stable or nearly stable. In other words, we're not going to make a W boson accelerator, because W bosons decay far too quickly. Likewise, we don't want to use tau leptons, or top quarks, or even heavy hadrons.

- Furthermore, you want something that is electrically charged. You wouldn't want to build a neutron accelerator for the simple reason that it's very difficult to accelerate neutrons. The way we accelerate particles in an accelerator is by grabbing onto them with electromagnetic fields, and electromagnetic fields will only accelerate electrically charged particles. So, we're not going to use neutrons, neutrinos, photons, or gravitons.

- That leaves us with electrons and positrons—and protons and antiprotons. We also have atomic nuclei. There are stable particles that are, for example, the nucleus of an iron atom or a lead atom. These are collections of protons and neutrons, but they are stable, and we can put them inside a particle accelerator.

- Muons are a possible thing that we could use in a particle accelerator. Besides electrons, positrons, protons, and nuclei, muons are on the edge of maybe being useful in a particle accelerator. They're unstable, and they do decay, but their lifetime is just long enough to make them an intriguing possibility. A muon collider is something that we might someday build.

- Given the particles we want to accelerate, how are we going to do it? Electric fields and magnetic fields are what we use to push around particles. We use electric fields to accelerate the particles, to get them moving quickly. We use magnetic fields to steer the particles, move the particles, or focus the beam of particles into a thin, powerful, highly intense column.

- In a real particle physics accelerator, the way that we get the charged particles to be boosted to higher and higher energies is something called a radio frequency cavity. The actual particles arranged in the accelerator are not smoothly distributed in the beam; they are concentrated in bunches with several particles each moving through. They are exquisitely timed, so when the bunch passes through the cavity, it is always feeling the electric field pushing it in the direction to get it to go to higher and higher levels of acceleration.

- In the parts when the electric field would be decelerating it, there are no particles there because that's in between the bunches. Through this technique, we can get the particles to come through grouped into bunches to be accelerated to faster and faster energies, very close to the speed of light.

Types of Accelerators

- We can accelerate particles along radio frequency cavities, but what do we do once we've nudged the particles to a high energy? There are two main geometries that we have for particle accelerators. A linear accelerator accelerates particles once and then collides them. With a circular accelerator, you can recycle the particles.

- Both circular accelerator`s and linear accelerators have their uses; there are trade-offs to using each one. Generally, we use circular accelerators when we have heavy particles, such as protons or antiprotons, and we use linear accelerators when we have light particles, such as electrons and positrons.

Notable Particle Accelerators

Early Cyclotrons	1930s	Berkeley	9–60 in	circular	Hydrogen ions	1–16 MeV	
Intersecting Storage Rings	1971–1984	CERN	948 m	circular	proton-proton	30 GeV	
SPEAR	1972–	SLAC, California	250 m	circular	e⁺/e⁻	7 GeV	charm quark, tau lepton
Super Proton Synchrotron	1981–1984	CERN	6.9 km	circular	proton-antiproton	300 GeV	W/Z bosons
Stanford Linear Collider*	1988–1998	SLAC, California	3 km	linear	e⁺/e⁻	90 GeV	
Large Electron Positron Collider*	1989–2000	CERN	27 km	circular	e⁺/e⁻	200 GeV	
Tevatron*	1983–2001	Fermilab, Illinois	6.3 km	circular	proton-antiproton	1 TeV	top quark
Superconducting Super Collider*	canc. 1993	Waxahachie, Texas	87 km	circular	proton-proton	40 TeV	
Large Hadron Collider	2008–	CERN	27 km	circular	proton-proton	7/8/14 TeV	Higgs boson
International Linear Collider?	2026?	Japan?	30–50 km	linear	e⁺/e⁻	500 GeV–1 TeV	

© Sean Carroll.

*really tried to find the Higgs boson

- If we're using protons, antiprotons, electrons, and positrons, do we want to collide particles with their antiparticles or with themselves? Again, there are trade-offs for each method. If you collide a particle with an antiparticle, you generally get a more efficient annihilation into whatever new particles you're trying to make, which is a good thing. The problem is that it's difficult to make antiparticles in the first place.

- A linear collider generally uses electrons and positrons. Positrons are fairly light, but because their mass is low, they're relatively easy to make. A circular collider is typically going to use protons. At the LHC, we collide protons with other protons.

- Protons and antiprotons are big floppy bags of particles. We can raise them to very high energies with hadron colliders, but these colliders don't give very precise information. As a result, hadron colliders are useful for discovering new particles. In contrast, an electron-positron collider cannot reach as high of an energy level, but the energy it reaches you can study with very high precision. So, once you've discovered something at a hadron collider, you can study it at an electron-positron collider.

- At CERN, the tunnel where the LHC is now built used to house the Large Electron-Positron (LEP) collider, which was a tool for exploring and studying a great number of the properties of all the particles in the standard model. Many of our best pieces of information about the standard model came from the LEP.

Particle Detectors
- How do you know what you've created by smashing particles together in a particle accelerator? To find out, you need to build a detector, which is a multifaceted instrument that looks for all the different kinds of particles that can be created.

- The particles that we're trying to create and discover, such as the Higgs boson or dark matter, are not the particles we will ever directly see in our detector. Instead, we're going to see the decay products. We're going to make a particle that will then produce the particles of the standard model. So, we're really trying to detect standard model particles.

- You're never going to detect some of the standard model particles, such as neutrinos or gravitons, because they're just not interacting strongly enough to be detectable. They will fly out of your detector, so you might as well ignore them.

- Some particles—such as the Higgs boson, the tau, the top quark, and the W and Z bosons—decay so quickly that you'll never see them directly. You're just going to see what they turn into.

- You could make strongly interacting particles, such as quarks or gluons, in a particle physics collision, but these particles are confined. If you make an individual quark or gluon, it scatters into a whole bunch of strongly interacting particles. This process is called hadronization, because strongly interacting particles are hadrons.

- The collection of all these hadrons that spill out in a certain direction is called a jet of particles. It's very difficult to separate out the different hadrons inside a jet, but we collect all of the energy in every individual jet and put it into the information that we get out of the decay.

- Besides hadrons, that leaves us with the electrons (and the anti-particles, the positrons), the photons, and the muons. This is the collection of particles we might want to detect in our detectors. Electrons and photons can be absorbed. Hadrons are heavy but interact strongly, so they can also be absorbed. It's only muons that can punch through lots of material and yet be detected but not absorbed.

- With these considerations in mind, we can build a detector. A modern particle physics detector comes in pieces, because there are different particles that can come out of the interactions, and there are different ways we want to measure those particles.

- In general, any modern particle physics experiment is going to have four layers.
  - At the center is an inner detector that offers a very high-precision view of what's happening with precision measurement and timing.

  - The next layer around the inner detector is called the electromagnetic calorimeter, which captures the particles and measure their energy as well as their momentum.

  - Around the electromagnetic calorimeter is the hadronic calorimeter, which captures the jets of the strongly interacting protons and also measures the spray of energies inside the jets.

  - Finally, around the hadronic calorimeter is the muon detector. This outer layer cannot capture the muons, but it can measure their path and, therefore, figure out their momentum, or energy.

- From all of this information we can reconstruct, at least probabilistically, what kind of event would have created all of these particle tracks.

Suggested Reading

Carroll, *The Particle at the End of the Universe*, Chapter 6.

Kane, *The Particle Garden*, Chapter 5.

Lederman, *The God Particle*, Chapter 6.

Questions to Consider

1. The LHC experiments throw away most of the data they produce. How would you set up the triggers that decide what to keep and what to throw away?

2. Can you think of ways to discover new heavy particles, other than to smash known particles together?

The Large Hadron Collider
Lecture 9

In this lecture, you will learn about the process of building the machine that discovered the Higgs boson: the Large Hadron Collider. It is the most complicated machine ever built by human beings. By design, no one was trained to build this machine, because nothing like it had ever been built before. The people who built the LHC had built other particle accelerators and had been involved in the design of the LHC, but when you build something of this scope and magnitude, there will always be unforeseen challenges. A lot of human wisdom, ingenuity, and gumption went into achieving this incredible feat.

CERN and the Birth of the LHC

- Established in 1954, CERN is the European center for nuclear research. These days, most of the research concerns particle physics, not nuclear physics. Currently, there are 21 countries that govern the operation and contribute most of the budget, called member states, but there are 70 countries that participate in many ways, called observer nations. CERN is a model of international cooperation.

- In 1983, CERN discovered the W boson and the Z boson, and they won the Nobel Prize the next year, in 1984. Specifically, the Nobel Prize went to Carlo Rubbia, who was the leader of an experiment at CERN called UA1. He shared the Nobel Prize with another physicist at CERN named Simon van der Meer, who received it for inventing a technique known as stochastic cooling, which made modern particle accelerators possible.

- After Rubbia found the W and Z bosons, particle physicists wanted to find the next thing. In the early 1980s, there were two very obvious candidates: the Higgs boson and the top quark. In 1995, the top quark was discovered at Fermilab by the Tevatron. But CERN was looking for it, and they definitely wanted to find the Higgs boson.

- The next project in the pipeline for CERN was the Large Electron-Positron (LEP) collider, which ran from 1989 to 2000 and was certainly looking for the Higgs boson. Colliding electrons and positrons, it created Z and W bosons at an enormous rate, but it could not quite find the Higgs.

- The heaviest mass that could be reached at the LEP was about 114 GeV. There were hints that there was a Higgs boson at about 115 GeV—just beyond the reach of the LEP. CERN ran the LEP for a few months past its expiration date and then decided to turn it off. They did not find the Higgs at 115 GeV.

- Meanwhile, the United States had the Tevatron at Fermilab and was planning to build the Superconducting Super Collider (SSC), which would have reached 40 trillion electron-volts (TeV) of energy. The Europeans halted building their own 40 TeV machine.

- The SSC was canceled by Congress in 1993, at which point Rubbia argued that CERN should design its own hadron collider, named the Large Hadron Collider (LHC). Welsh physicist Lyn Evans guided the LHC from conception to design, to building, to ultimate operation.

- The target for the LHC was 14 TeV of energy, which was quite a bit higher than the previous hadron collider at CERN, the LEP. The SSC would have been 40 trillion TeV, but the LHC was constrained because they were saving money by building it inside the same tunnel that the LEP had been built in.

How the LHC Operates
- The LHC was built in a 27-kilometer (100-meter) ring underground, beneath the French-Swiss border. It is a tunnel that is 27 kilometers around with some of the highest tech, most expensive equipment that human beings have ever made.

- Most of that length consists of superconducting magnets to collimate, or focus, the beams of protons to a very tiny amount and to steer them around the ring. There are about 1600 superconducting magnets linking together underneath the ring, and each magnet weighs 27 tons.

- The actual acceleration of the particles is done by a radio frequency cavity, which takes up one small segment of the LHC tunnel. The rest of the ring is dedicated to steering and focusing the protons.

- The superconducting magnets steer the beams of protons. There are two beams, with protons moving in opposite directions. Inside the tubes, it is almost empty space; the atmospheric pressure inside these vacuum tubes is less than you would feel if you were standing on the surface of the Moon. They're also extremely cold, because the magnets themselves are superconducting.

- Through the magnets, protons are moving at a speed of about 99.99999% of the speed of light. The target energy in the beam is about 14 TeV total. There are about 600 trillion protons in the beams at any one moment in time. One gram of protons has over a trillion trillion protons, so there are not many protons by weight in the LHC beam at any one moment.

- The protons are all collimated to a very thin beam that is about two thousandths of a centimeter across. When these two beams collide, there are about a billion collisions per second. Every one time a bunch crosses, there are tens of collisions.

- This results in a lot of data, and a lot of it has to be thrown away through a trigger. But even once all but one in a million events is thrown away, we're left with about 30 petabytes of data every year. A peta is a million billion, so a petabyte is a million billion bytes. Every year at the LHC, we make five times the biggest database that we've ever had on Earth before that.

- All of these gigantic numbers come from one tiny little proton source, which has many, many hydrogen atoms in it. We strip off the electrons from the hydrogen atoms, and then we get protons and send them around the beam.

- Once the protons are moving around the ring in both directions, at certain points around the ring the beams of protons will be brought into collision. That's where we want to build a detector, which will watch what happens when the individual particles smash together.

- At the LHC, there are four large detectors, along with several smaller ones. Of the four large ones, two of them are general-purpose detectors, meaning that they look for everything, and two of them are more specialized. The general-purpose ones are called ATLAS and the Compact Muon Solenoid (CMS).

- ATLAS is 25 meters across, 46 meters long, and weighs 7000 tons. CMS is over 21 meters long, 15 meters across, and weighs over 12,000 tons. CMS is smaller in size than ATLAS, but it's heavier, because CMS is much more densely packed.

- In addition to the two major experiments, ATLAS and CMS, there are the dedicated experiments, ALICE and LHCb. Instead of looking at protons colliding, ALICE looks at lead ions colliding—specifically, the nuclei of lead atoms. By colliding together these tremendous collections of protons and neutrons, we create the quark-gluon plasma that mimics the early universe. We are recreating the conditions from the big bang to understand the physics underlying the interactions of many, many quarks and gluons.

- Finally, the LHCb ("b" stands for "beauty") detector is looking for CP violation using bottom, or beauty, quarks. In the term "CP violation," "C" stands for "charge conjugation," which is the symmetry, or at least the transformation, that exchanges particles for antiparticles. "P" stands for "parity," which is violated in the weak interactions. The weak interactions violate parity, but they come very close to conserving the combination of parity and charge conjugation, which is called CP. The LHCb provides an indirect way of discovering new particles—by measuring CP violation.

The Life of the LHC
- In 1984, we started imagining building the LHC. In 1993, construction was started in earnest. On September 10, 2008, it was finally turned on. On September 19, nine days later, part of the machine exploded. Six tons of liquid helium were splashed across the floor. All of the control room computers lit up bright red. Lyn Evans, the boss of the LHC, described it as carnage.

- What happened is something called a quench. One of the superconducting magnets, which should be kept cold, heated up and exploded. It took over a year to fix things, but in November of 2009, the LHC came roaring back to life. But they ran it at only seven TeV, which is half of the targeted total energy, to be cautious. Since November of 2009, the LHC has been working very well.

- The LHC was running at seven TeV for a long time. Then, in 2012, it went up to 8 TeV, which is exactly what was needed to finally discover the Higgs boson. The Higgs boson was discovered in July of 2012. The discovery was announced, and the LHC ran for a few more months. But at the end of 2012, it was shut down.

- Throughout 2013 and 2014, the LHC was not running, because it was being upgraded. When it turns back on in 2015, the LHC is going to run at its design energy—14 TeV. We're hoping that the extra energy is what we need to discover particles beyond the Higgs boson, beyond the standard model of particle physics.

Suggested Reading

Butterworth, *Most Wanted Particle*, Chapter 2.

Carroll, *The Particle at the End of the Universe*, Chapter 5.

Randall, *Knocking on Heaven's Door*, Chapter 13.

1. How do you think the practice of physics is changing in the Big Science era, when experimental collaborations sometimes involve thousands of people?

2. What arguments would you make—pro or con—regarding a new proposed particle accelerator that would reach much higher energies but would be relatively expensive?

Capturing the Higgs Boson
Lecture 10

O nce we've built the machine that is needed to find the Higgs boson, the Large Hadron Collider, we first need to make the new particle. Then, we need to detect it. However, we won't ever see the Higgs boson directly. If we were to make it, the Higgs boson would decay very, very quickly—in one zeptosecond (10^{-21} seconds), to be exact. Therefore, we're actually looking for the particles that it decays into. Finally, we have to know that what we're seeing did, in fact, come out of the decay of the Higgs boson.

Making the Higgs Boson

- With a hadron collider, we are colliding two protons together. Protons have two up quarks and one down quark inside—three quarks total. The strong interactions that keep those quarks together are very strong, resulting in a tremendous number of virtual particles inside the proton. There are many gluons that are holding the three quarks together. There are also quark/antiquark pairs popping in and out of existence.

- When you collide these two protons together, it's like you're colliding two floppy bags of particles, and you're not going to get the total energy in your proton to be concentrated in any one collision. When you're colliding particles at the Large Hadron Collider (LHC) at 7 trillion electron-volts (TeV) total (less than the originally scheduled 14 TeV), that's 3.5 electron-volts per proton. But at most, one-third of that is in any one quark, and often it's much less.

- We still have plenty of energy to make a Higgs boson. When we collide two protons, it's either going to be two quarks colliding, or a quark and an antiquark, or two gluons. We want the total energy in those individual constituents to be enough to make one Higgs boson and then whatever other particles might be produced in that collision.

- When the LHC turned on in 2005, we had already been looking for the Higgs boson for quite some time. The LEP accelerator at CERN, along with accelerators at SLAC. Fermilab, and elsewhere, had ruled out quite a wide range of possibility.

- The only thing we didn't know about the Higgs before we discovered it was its mass. Once we knew its mass, we would instantly know how it interacted with all the other particles in the standard model. So, in 2008, there were basically two possible windows for the Higgs: one was between 115 and 155 GeV and the other was greater than 250 GeV. It turns out that the Higgs was found at 125 GeV.

- How do we find the Higgs? What particles does it couple to? We want to both produce it and then watch it decay. The Higgs boson couples to every particle that has a mass—and the bigger the mass, the stronger the coupling. Therefore, the particles that the Higgs couples to the most would be the W and Z bosons and the very heavy fermions.

- When thinking about making the Higgs boson, we have our ingredients, or our protons. which are made of three quarks. Inside the proton there are virtual particles, including quark/antiquark pairs and gluons bouncing back and forth between quarks.

- But the gluons and quarks that make up the proton—the up quarks and the down quarks—are very light particles. So, the particles that are inside the proton are not very good for making Higgs bosons. In fact, the math proves that the best way to make a Higgs boson by proton collisions is to first turn those light quarks and gluons into heavy particles, such as W and Z bosons or top works, and then use those to make the Higgs.

- In a proton-proton collision, what is the easiest way to make these heavy particles—top quarks, W bosons, and Z bosons? There are two ways: via the strong interactions or the weak interactions.

- One way to make a Higgs boson is by first making a top quark. It is easiest to make the top quark through the strong interactions; all quarks feel the strong interactions. Basically, you want to take gluons and convert them into top quarks, and then let those top quarks make the Higgs boson. Or you can make a top/anti-top pair that annihilates into a Higgs as well as making other strongly interacting particles.

- The other way is by first making a W or Z boson pair. To do that, you use the weak interactions. You bring together the quarks that you already have in your proton—the up quark, the down quark, etc.— which couple to the W and Z bosons via the weak interaction. Then, there are two ways to make a Higgs boson: One way is to combine a positive W and a negative W or two Z bosons to make a Higgs, called fusion. The other way is emission. You can emit from the quarks a W or a Z boson, and then they can emit in turn a Higgs boson.

Decay and Theoretical Predictions

- When we use Feynman diagrams to calculate transition probabilities, keep in mind that they are probabilities. We don't say for sure that a certain interaction will lead to a Higgs boson. If you were to make a Higgs boson, a certain fraction of the time it will decay into some particles, and other times, it will decay into something else. That's just how quantum mechanics works.

- Once we've made the Higgs boson, we're going to calculate the probability for it to decay into something else. The Higgs wants to interact with heavy particles. Therefore, it wants to decay into heavy particles. However, if the particle is too heavy, the Higgs doesn't have enough energy to decay into it.

- In fact, a Higgs of 125 GeV can "decay" into particles that are heavier than it, because the Higgs itself can be a virtual particle. And its mass is not actually its real mass. However, if the mass of a virtual particle is far away from the mass of the real particle, that gives you a very small contribution. The Higgs wants to decay into particle/anti-particle pairs that are similar in mass to the Higgs itself.

- The Higgs does not couple to photons, which are neutral and massless. However, we can't ignore the photons. The Higgs can decay into heavy particles, such as top quarks, which couple to photons. So, you can go through a virtual heavy charged particle.

- In fact, when we actually found the Higgs, the strongest signal we had was from the Higgs decaying into two photons via some virtual heavy charged particle. That's not because it happens most of the time; it's just a much cleaner thing to happen. There are not many other ways of making two very energetic photons.

- But it's not the only thing that can happen. There is a whole list of things that can happen. For example, the Higgs can decay into two Z bosons. Likewise, the Higgs can decay into two virtual W bosons, because it's not quite heavy enough to make two real W bosons. Then, the W bosons decay. But because the W bosons are charged particles, there are fewer ways for them to decay than for the Z boson.

- Another possibility is that the Higgs boson can decay directly into quarks. This is something the Higgs likes to do more than anything else, because quarks are heavy. So, the Higgs can decay into top quarks, which are virtual, or even bottom quarks, which the Higgs can make for real.

- Any particle that decays, including the Higgs boson, likes to decay into real particles rather than virtual ones. So, bottom quarks are the favorite thing for the Higgs boson to decay into. Unfortunately, we don't observe bottom quarks; we observe what they decay into. In our detectors, we're looking for jets of strongly interacting particles—jets of hadrons. In particular, we're looking for two jets, because the Higgs creates a quark/antiquark pair, each of which makes a jet.

- Finally, the Higgs boson decays into tau leptons. The tau is the heaviest lepton, so the Higgs can decay directly into it. Then, the tau itself decays very quickly into an electron, a muon, or directly into quarks and hadrons, making a jet. There are also invisible neutrinos, which are a problem because they carry away energy. Therefore, we can't measure the total energy produced in the event.

- If we're looking at a Higgs decay, where every part of the decay product is visible, then we can add up all of the energies and get the mass of the Higgs boson. If a lot of the energy is carried away by neutrinos, then we are less sure of the result of finding a Higgs boson.

- When the Higgs decays, over half of the time it will be into a bottom/anti-bottom pair. But the other possibilities—including W pairs, gluons, and tau/anti-tau—also exist. Because we don't observe the Higgs directly, we would like to see that this new particle we detect makes all of the possibilities in the right proportions by compare all of the data points to our theoretical predictions.

Detecting the Higgs Boson

- Let's say that you observe an event with a detector like ATLAS or CMS. You've taken a picture of the spray of particles being emitted once the two protons have collided. You get some low-energy particles that are not that important, and you get four particles that have a lot of energy: an electron, a positron, a muon, and an anti-muon. This is what you might expect if you had first made a Higgs boson and that Higgs boson had decayed into two Z bosons and those Z bosons had decayed into lepton/anti-lepton pairs.

- You've seen this event, and it fits in with your expectation from the Higgs. But is it the Higgs, or is it something else? The problem is that every Higgs decay can be produced by some non-Higgs mechanism. Higgs decay is an extra way to make those particles, especially with total energy close to the Higgs mass—125 GeV.

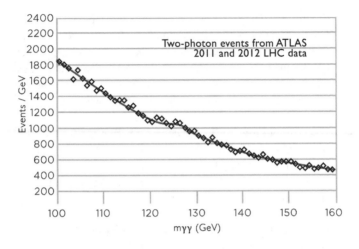

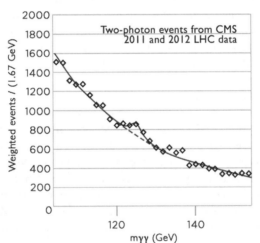

Adapted from "Observation of a new particle in the search for the Standard Model Higgs boson with the ATLAS detector at the LHC" by ATLAS Collaboration, Physics Letters B Volume 716, Issue 1, 17 September 2012, Pages 1–29. Creative Commons Attribution License 3.0

- So, you're looking for a small amount of excess in the number of certain kinds of particles produced with a certain characteristic energy. If you were to plot all of the particles that you're observing as a function of the total energy in those particles, what you're predicting—because there is a Higgs boson—is a little bump in the data.

- If you see a bump in your data, you want to make sure that the bump is evidence that there is a Higgs boson and not that it is just a random fluctuation. This is verified by conducting a statistical analysis.

- On December 13, 2011, there were two seminars at CERN by Fabiola Gianotti and Guido Tonelli, the spokespeople for ATLAS and CMS at the time, and evidence was presented for the Higgs boson. They both saw bumps in their data, and the bumps were statistically significant.

- It wasn't until July 2012 that we actually claimed to have discovered the Higgs. But what we really knew in then was that we had found a new particle. There was a statistically significant bump in the data, but how did we know it was the Higgs boson?

- We had to keep studying this new particle. We studied the Higgs decay into two photons, into quarks, into leptons, and into everything else we could find. It's the fact that all of these measurements fit into a single theory that gave us confidence that we were on the right track.

- In addition, we measured the spin and the parity of the new particle, along with the relative amount of coupling of this new particle to everything else. Whatever this new particle was, it was coupling preferentially to heavy particles, not to light particles.

- On July 4, 2012, CERN announced that they discovered a new particle. Then, in March of 2013, the scientists from ATLAS and CMS said that the new particle was definitely the Higgs boson. Soon after, the 2013 Nobel Prize went to Francois Englert, Peter Higgs, and Robert Brout.

Suggested Reading

Butterworth, *Most Wanted Particle*, Chapter 5.

Carroll, *The Particle at the End of the Universe*, Chapter 9.

1. How well do you think people understand and appreciate the idea of statistical significance in their everyday lives? Do you think it has wide applicability?

2. Is the Nobel Prize a good idea, or is it a distraction from the pursuit of scientific truth? Should the traditions be changed so that more than three people can win it in one year?

Beyond the Standard Model of Particle Physics
Lecture 11

As exciting as the Higgs boson discovery was, the hope is that it is a bridge to new, unexpected physics. There are properties of the Higgs boson that hint at the idea that the physics we currently have is incomplete. In addition, we can use the Higgs boson itself as a window into finding new physics. A few ideas that motivate us to explore beyond the physics we already know are the existence of dark matter and dark energy, which is not explained by the standard model, and the existence of gravity, which has not been firmly established by quantum mechanics.

Looking beyond the Standard Model

- We found the last particle of the standard model, the Higgs boson, in 2012. But it was predicted long before that. And the last prediction that became part of the standard model was predicted in the 1970s. Since then, we've had a lot of time to try to move beyond the standard model.

- There have been many ideas that are inspired by the fact that the standard model can't be the whole story.
  - One idea is called grand unification. Weinberg, Salam, and Glashow showed us how the electromagnetic force is unified with the weak nuclear force, but there is still the strong force out there that is not unified with anything. Grand unification is the hypothetical idea that there is one big force that gives rise to the strong, the weak, and the electromagnetic interactions.

  - In addition, there are cosmological ideas, such as inflation in the early universe. Inflation requires new particles; you cannot get inflation out of the particles we know about in the standard model.

o   Furthermore, there are three generations of particles in the standard model. Perhaps there is a fourth generation. We haven't yet found it, but there are ideas about partial or even completely new generations.

o   Finally, protons and neutrons are made of quarks and gluons, but is it possible that quarks and gluons are made of even smaller particles? Or are electrons or Higgs bosons made of a substructure of even smaller particles, which would be called preons? We don't have any evidence of this idea yet, but it is still on the table.

Exploring the Possibilities
- We're exploring many different ideas, and there are clues in the existing data. The most prominent example in particle physics is called the hierarchy problem: Why are the parameters that are attached to the Higgs boson so small in energy?

- There are only two parameters for the Higgs boson: the mass, which is 125 GeV, and the value that the Higgs field takes in empty space, which is 246 GeV. These numbers are similar to each other, meaning that if something is Higgs related, it's going to be in the range of 100 to 200 GeV. This is 100 or 200 times the mass of the proton, but it turns out to be very small compared to other scales of particle physics.

- If grand unification is true—if the strong force is unified with the weak force—then that will happen at a scale of somewhere near 10^{16} GeV. If gravity is included, then this number increases to 10^{18} GeV, which is on the Planck scale. These numbers are huge compared to the 10^2 GeV that characterizes the Higgs boson. Why is the Higgs boson, the weak scale of particle physics, so different from these large scales of gravity or grand unification?

- This is not just a problem that considers why these numbers are different. We can try to estimate how big the numbers should be. To calculate a process, such as two electrons scattering off of each other, we add up many different Feynman diagrams, and if the graphs have loops inside, then we add up all the different ways that momentum and energy can flow through those loops.

- This is true even for a particle just moving through space. For example, the Higgs boson field is always gently interacting with all of the fields around it, so if we're calculating the property of the Higgs boson at rest—which is simply the mass of the Higgs boson—we add together a number of Feynman diagrams, and when we do that, the predicted mass of the Higgs boson is infinity.

- We think this means that we get an infinitely big contribution to the Higgs boson from the virtual particles. But that's only because we're including energies that the virtual particles could have that go all the way up to infinity.

- Let's say that we shouldn't include energies higher than some very large energy scale, such as the Planck scale, the scale at which quantum gravity kicks in and space-time itself probably breaks down. Then, we make a very finite precise prediction for what the mass of the Higgs boson should be—at the Planck scale, 10^{18} GeV. But it's actually 125 GeV. That is the hierarchy problem. What keeps the Higgs scale so low?

Supersymmetry
- We don't know the solution to the hierarchy problem, but we do have some scenarios that might help. First, we can imagine that in those virtual particles that contribute to the mass of the Higgs, some of them contribute a positive amount and others contribute a negative amount. Perhaps, for reasons we don't understand, there is a cancellation between the positive and negative contributions to the Higgs mass.

- Another possibility is that instead of including contributions all the way up to the Planck scale, there is some reason to stop including new quantum corrections at a much lower energy scale. If you could somehow put an upper limit on those lower contributions, you could save the mass of the Higgs and solve the hierarchy problem.

- Both of these solutions have been implemented in very specific models, one of which is called supersymmetry. We think that gauge symmetries are the explanation for all of the forces we know in the universe. Supersymmetry is a new kind of symmetry that has not yet been discovered in nature between bosons and fermions.

- The hierarchy problem says that the Higgs boson should have a very big mass because it gets quantum corrections from virtual particles. Those additional contributions to the Higgs mass are positive from boson virtual particles and negative from fermion virtual particles.

- So, it's possible that they could cancel each other. But they would only cancel each other if the set of bosons in the universe that were contributing to the mass of the Higgs were matched, one to one, with a set of fermions that contribute to the mass of the Higgs. In the standard model, that's not true. But supersymmetry says that every boson has a matching fermion partner with the same properties—including mass, charge, and interactions—and vice versa, except spin.

- The problem with supersymmetry is that standard model particles don't pair up. This doesn't bother physicists very much because they know that when you invent symmetries, you can break them. For example, the Higgs boson breaks the symmetry of the electroweak theory. Maybe there's something that breaks supersymmetry and hides the superpartners from us.

- The prediction of supersymmetry is that for every particle we know about in the standard model, there is a partner with very different mass but with the same interactions, charge, and opposite spin. We call these "sparticles." If supersymmetry is right, there are many, many new particles to be discovered.

- How would supersymmetry work in the real world?
  - Supersymmetry solves the hierarchy problem if the superpartners are near the electroweak scale. This is promising for future experiments.

○ Sparticles don't necessarily match up to the original particles in an easy way because they can mix together.

○ Because of the mathematics of supersymmetry, there's a very strong, clear prediction that there are at least five different particles that play the role of the Higgs boson in the standard model—three neutral and two charged.

○ Supersymmetry is predicted by string theory but could exist without it.

- There are many reasons to take supersymmetry seriously, even though we haven't found it yet. It solves the hierarchy problem, is predicted by string theory, and gives us a very nice candidate for the particle that is the dark matter of the universe, plays a crucial role in the idea of grand unification by helping the coupling constants unify.

Alternatives to Supersymmetry

- Supersymmetry remains the favorite theory beyond the standard model among modern particle physicists, but there are alternative theories. The alternatives, for better or worse, are sort of falling by the wayside one by one. We are making progress in particle physics by ruling out some people's favorite ideas.

- For example, there's an idea called technicolor. Color is what goes into the strong interactions, which are sometimes called quantum chromodynamics. A quark has a color: red, green, or blue. The idea of technicolor is to add a new kind of color that is unrelated to the color that quarks have in quantum chromodynamics and call it technicolor.

- Technicolor would be a new strong force that is stronger than the strong force. It would confine its "techniquarks" together in very small packages. Then, you would predict a whole bunch of particles as bound states of these techniquarks. One of these could be a spin-0 boson that would be similar to the Higgs particle.

- In other words, in technicolor, the Higgs boson, or the thing that plays the role of the Higgs boson, is not a fundamental particle all by itself. It's a bound state of these new kinds of particles, these techniquarks that feel technicolor. This theory does fix the hierarchy problem. Those contributions to the mass of the Higgs don't affect a bound state of techniquarks in the same way that they would affect a fundamental particle. The problem is that technicolor models don't seem to fit the data.

- In the late 1990s, many theorists suggested that we could try to address the hierarchy problem using extra dimensions of space. The idea that space has more than the conventional three dimensions is an old idea. But even if there are extra dimensions, we have experimental evidence that they must be at least less than about 100th the size of the proton, or less than 10^{-17} centimeters.

- What if the extra dimensions are large, and the reason we don't see them is simply because we can't get there? In other words, imagine large extra dimensions of space, but there's a three-dimensional surface embedded in these dimensions—a surface we call a brane. If all of the particles and forces of the standard model were confined to this three-dimensional brane, then space would look three-dimensional to us, even if there were larger extra dimensions we didn't see.

- But gravity always escapes to the extra dimensions. This means that the extra dimensions can't be too big. Therefore, gravity seems weaker than it really is to us. We've only tested gravity at distance scales of about one millimeter across. It's only because gravity is diluted in the extra dimensions that it seems weak to us.

- This is an approach to solving the hierarchy problem by saying that there's not a hierarchy—there's no such thing as the Planck scale, and all of the new physics in the world in this scenario is at about 100 or 1000 GeV, where we're doing experiments today.

- Our tests of gravity have improved; we have now tested it down to 10^{-2} centimeters. We know that gravity does not break down at 1 millimeter. This theory predicts that gravity becomes strong at the TeV scale. Therefore, you can actually make gravitons. However, we don't see them.

- The idea that extra dimensions were large and could help solve the hierarchy problem was a very good one, but it seems not to have worked. Supersymmetry is still the leading candidate for a theory beyond the standard model. But it could have been found and hasn't been. However, we only just found the Higgs boson. There is plenty of room for discovery, and we should be open to surprises.

Suggested Reading

Carroll, *The Particle at the End of the Universe*, Chapter 12.

Kane, *The Particle Garden*, Chapters 9–10.

Questions to Consider

1. What do you think the chances are that a fine-tuning such as the hierarchy problem is just an accident and has no deeper explanation?

2. How should physicists balance the search for new models that have been proposed by theorists against looking for complete surprises that nobody has anticipated?

Frontiers—Higgs in Space
Lecture 12

We have a very good model that explains the physics on Earth: the standard model of particle physics. And we have just discovered the final piece of the Higgs boson. But our theory is incomplete. We live in a universe that is made up of about 95% dark matter and dark energy, which are not explained by the standard model. This final lecture will consider the possibility that both dark matter and dark energy might be related to the Higgs boson—that we might be able to use our discovery of the Higgs to learn more about the other 95% of the universe.

Dark Matter
- Many people have doubted that there is dark matter. But dark matter exists. The first hints of dark matter go back to the 1930s with astronomer Fritz Zwicky, who observed clusters of galaxies, and it became a solid case in the 1970s with Vera Rubin and Ken Ford, who showed that there was a lot more gravitational pull at the edges of galaxies than you can account for with the stars, gas, and dust inside.

- Today, we have more data that tells us so much about the dark matter that we know it cannot be ordinary matter. For example, there is gravitational lensing. Every bit of matter in the universe affects the gravitational field, and gravity, in turn, deflects light as it passes through. We can use gravitational lensing to map out where the dark matter is in the universe.

- There's also the evolution of large-scale structure in the universe, which is clearly being influenced by something other than ordinary matter. Finally, there's the cosmic microwave background, the leftover radiation from the big bang. The subtle patterns of the temperature of the microwave background depend on what the universe is made up of. Using only ordinary matter, you don't get the right prediction for what the microwave background looks like, but dark matter and dark energy fit perfectly.

- There's a lot more dark matter in the universe than there is ordinary matter. Perhaps the most vivid illustration of this is the Bullet cluster, which is two clusters of galaxies that pass through each other. In a cluster of galaxies, most of the ordinary matter isn't in the galaxies, and we think that most of the matter overall is in the dark matter in between the galaxies. Data shows that most of the mass in the Bullet cluster is not where most of the ordinary matter is.

- How do we know that the dark matter isn't simply ordinary matter that is dark? We have separate ways of talking about and measuring the amount of ordinary matter in the universe. It's not just counting the stars, the gas, and the dust. Depending on how many ordinary particles there are in the universe, it affects the rate at which particles come together by nuclear fusion to create new elements in the early universe.

- For example, in the early universe, hydrogen and helium fused together to make lithium and other elements, including deuterium. And the rate at which this happens depends on how much ordinary matter there is. Likewise, the patterns in the cosmic microwave background are closely affected by the ratio of the number of baryons, of protons and neutrons, to the number of photons in the universe.

- And in both cases, we get the same answer: Only 5% of the energy density of the universe can be explained by the standard model. We need a new particle—a dark matter particle. It must be stable, neutral, and hardly interacting.

- What is the dark matter? We don't know, but there are two candidates for dark matter that stand up above the rest: a weakly interacting massive particle (WIMP) and an axion particle, both of which are related to the Higgs boson. In fact, the axion is a lighter cousin of the Higgs boson.

- The WIMP is the most popular candidate for what the dark matter is. Think about the early universe, in which the density of matter was much higher, and so was the temperature. Particles and antiparticles, including dark matter, were being created or destroyed. But as the universe expanded and cooled, most of these particles and antiparticles annihilated with each other and disappeared into photons.

- After the density decreased so much that the particles couldn't annihilate anymore, there was a leftover abundance of particles. This is the relic abundance that gives us, in principle, the dark matter abundance today.

- What is the rate of interactions between the dark matter particles and antiparticles that you would need to get to give us the correct abundance of dark matter today? The answer is the dark matter particles need to interact with a strength that is exactly the strength of the weak interactions. This is known as the WIMP miracle. If the dark matter is a particle that interacts through the weak interactions, it's very easy to get the right abundance of dark matter particles.

- The dark matter particle must interact with either Z bosons or Higgs bosons via annihilation. But if a dark matter particle and an antiparticle annihilate into a Higgs boson, then that Higgs boson can decay into particles in the standard model. If it is true that the dark matter feels the weak interactions, or interacts with the Higgs directly, that means that the dark matter interacts with ordinary matter directly as well. And that means that we can detect the dark matter.

- The way to detect the dark matter is to build a detector deep underground that has very heavy nuclei in it, because a nucleus has protons and neutrons. The WIMP will scatter off individual protons and neutrons. Then, a WIMP dark matter particle scatters off the heavy nucleus, shaking the nucleus and, in turn, the electrons in the atom. And that emits light. This is the way we will ultimately try to detect dark matter.

- In fact, this is exactly what we're doing. There are several different dark matter experiments going on right now. For example, in the Black Hills of South Dakota is the LUX detector, which is 350 kilograms of liquid xenon, which is a heavy element that is neutral and doesn't interact very happily with ordinary matter.

- We've made a tremendous finding in our search for dark matter. The data has ruled out the possibility that there are dark matter WIMP particles in the range we're thinking about that are scattering off ordinary particles by exchanging Z bosons. However, we have not ruled out the possibility that dark matter could scatter off ordinary matter by exchanging Higgs bosons. If the dark matter is a WIMP, it interacts with ordinary matter via exchanging Higgs bosons. That is a feature of the Higgs.

- The Higgs is a sociable particle. It is easier for it to interact with new particles, so whether or not we actually find dark matter by this technique, the fact that we found the Higgs boson gives us new hope for finding new particles beyond what we know in the standard model.

Dark Energy
- We don't know as much about dark energy as we know about dark matter, which is very little. We do know that the dark energy is approximately the same density of energy at every location in space and over every moment of time. That's why we know that dark energy is not a particle. Particles dilute as the universe expands.

- The dark energy could be the intrinsic energy of empty space. It could be the idea that in every cubic centimeter of space, you can remove everything there is, including all of the ordinary matter and the dark matter, so that it's nothing but a vacuum, or empty space.

- How much energy is there in that cubic centimeter? According to Einstein, the answer is not necessarily zero. It's some constant of nature—the vacuum energy or the cosmological constant. That could be the dark energy.

- Think about this in terms of the Higgs field. The reason why the Higgs boson takes a nonzero value in empty space is because of its potential energy. The potential energy function for the Higgs boson is simply the answer to the following question: How much energy is there in empty space when the Higgs boson field takes on different values?

- The minimum energy is when the Higgs is at a value of 246 GeV. But what is the actual value of the energy? The answer is that it's completely unknown. We can have any energy we want at the minimum value. That is the dark energy. That is the vacuum energy that is apparently making the universe accelerate.

- We now know the complete Higgs boson potential. But just like everything else in particle physics, the Higgs potential gets corrections from quantum mechanics. The good news is that once we have measured the mass of the Higgs and its value in empty space, which we have done, we can calculate all of the quantum corrections. So, we can plot the complete and final form of the Higgs boson potential.

- And the answer is fascinating. There is a minimum value for the Higgs potential energy at its current value, 246 GeV. But there is another minimum value very far away at large values of the Higgs field.

- Even though right now the Higgs field takes on the value of 246 GeV, it might change. There is something called quantum tunneling. You can start with the Higgs field having its current value, but in a tiny region of space, a bubble can form. Through a quantum fluctuation, you could make a tiny region where the Higgs field takes on a much larger value, but with a lower energy.

- And that has dramatic consequences. If you have some energy in some region, it wants to minimize, or spread out. Therefore, that bubble that forms will want to grow, because that is lowering the energy of space itself. There is a competition: The universe is expanding, but bubbles are being created.

- Once a bubble is created, it starts expanding at almost the speed of light. Is the rate at which these bubbles come into existence fast enough that the bubbles can collide and percolate? Percolation is a phase transition much like liquid water boiling and becoming water vapor by creating bubbles. If you create bubbles of new Higgs value fast enough, the bubbles will hit each other, and take over the universe, and we will have a phase transition that is percolated with a new value of the Higgs.

- What would that mean for us? Life would not exist. All the energy in empty space would convert to Higgs bosons, which would then decay. Everything that we know about chemistry, biology, and atomic physics would be wrong. Everything that is made of ordinary matter would explode and collapse into black holes.

- Fortunately, this hasn't happened yet, and it might not ever happen. But what matters is a delicate calculation: How fast do such bubbles form? Now that we know the mass of the Higgs, we can calculate the rate at which such bubbles of new Higgs value might appear in our universe.

- Our universe is about 10 billion years old, or about 10 billion light-years across in all three dimensions of space. We think that so far there has been no nucleation of bubbles in our universe that we know about. Whatever the nucleation rate is, it's probably less than one per 10 billion cubic light-years per 10 billion years. And in the real world, it's probably smaller than that. Based on the calculations, it is impossible for us to tell whether we will undergo this phase transition in the future.

Suggested Reading

Giudice, *A Zeptospace Odyssey*, Chapter 12.

Panek, *The 4 Percent Universe*, Chapters 7–9.

Randall, *Knocking on Heaven's Door*, Chapter 21.

Questions to Consider

1. Is it surprising that ordinary matter is a small fraction of the universe, or is it just what we should have expected all along?

2. Does the possible decay of our universe seem like a disturbing possibility, even if it's billions of years in the future? Which possible future for the universe seems likely to you: perpetual expansion and cooling, coming to an end at a "big crunch" (an infinite series of expansion and contraction), or constant change in unpredictable ways?

Bibliography

Butterworth, Jon. *Most Wanted Particle: The Inside Story of the Hunt for the Higgs, the Heart and Future of Physics.* New York: The Experiment, 2015. A lively recounting of the building of the LHC and the discovery of the Higgs by a writer who is also a professional physicist on the ATLAS experiment.

Carroll, Sean. *The Particle at the End of the Universe: How the Hunt for the Higgs Boson Leads Us to the Edge of a New World.* New York: Penguin Group, 2012. Similar in spirit and tone to this course, this book takes us up to the moment of the Higgs discovery and includes more explanation of the basic rules of quantum field theory and particle physics.

Close, Frank. *The Infinity Puzzle: Quantum Field Theory and the Hunt for an Orderly Universe.* New York: Basic Books, 2011. A detailed history of the development of quantum field theory.

Crease, Robert P., and Charles C. Mann. *The Second Creation: Makers of the Revolution in Twentieth-Century Physics.* New York: Collier Books, 1986. A historical overview of the progress of modern particle physics.

Giudice, Gian. *A Zeptospace Odyssey: A Journey into the Physics of the LHC.* New York: Oxford University Press, 2010. A theoretical physicist at CERN explains the detailed workings of the standard model and the LHC.

Kane, Gordon. *The Particle Garden: The Universe as Understood by Particle Physicists.* New York: Perseus Books, 1995. A view of the standard model and its possible extensions by a leading particle theorist.

Lederman, Leon, and Dick Teresi. *The God Particle: If the Universe Is the Answer, What's the Question?* Boston: Houghton Mifflin, 2006. Despite the unfortunate title, this book is a compulsively readable and understandable account of particle physics in general and the Higgs in particular.

Panek, Richard. *The 4 Percent Universe: Dark Matter, Dark Energy, and the Race to Discover the Rest of Reality.* Boston: Mariner Books, 2011. A view of dark matter, dark energy, and the rest of the universe beyond the standard model of particle physics.

Randall, Lisa. *Knocking on Heaven's Door: How Physics and Scientific Thinking Illuminate the Universe and the Modern World.* New York: Ecco, 2011. A leading theoretical physicist explains both modern particle physics and how it fits into a broader context of science and culture.

Sample, Ian. *Massive: The Missing Particle That Sparked the Greatest Hunt in Science.* New York: Basic Books, 2010. A great book about the search for the Higgs, before it was actually discovered.

Strassler, Matt. *Of Particular Significance.* http://profmattstrassler.com/. An online blog by an active particle theorist explaining underlying ideas and new discoveries.

Wilczek, Frank. *The Lightness of Being: Mass, Ether, and the Unification of Forces.* New York: Basic Books, 2008. A short introduction to the principles underlying particle physics by a Nobel Prize–winning physicist.